SMART SOLUTIONS

Skills, Problem Solving, Tools, and Applications

Whole Numbers and Money

Jan Phillips

New Readers Press

Acknowledgments

Advisers to the Series

Connie Eichhorn
Supervisor of Transitional Services
Omaha Public Schools
Omaha, NE

Lois Kasper
Instructional Facilitator
N.Y. Board of Education
New York, NY

Jan Phillips
Assistant Professor
William Rainey Harper College
Palatine, IL

Mary B. Puleo
Assistant Director
Sarasota County Adult and
Community Education
Sarasota, FL

Margaret Rogers
Coordinator
San Juan Unified Adult Education
Sacramento, CA

Consultant/Field-tester

Sandra Paris
Instructional Facilitator
New York City Public Schools
Office of Adult and Continuing Education

Library of Congress Cataloging-in-Publication Data

Phillips, Jan, date.
Whole numbers and money / Jan Phillips.
p. cm. — (Math solutions)
ISBN 1-56420-118-X (pbk.)
1. Arithmetic. I. Title. II. Series.
QA115.P57   1995
513.2'1—dc20   95-18618

Note: The title of this series is now *Smart Solutions.*

ISBN 1-56420-118-X

Copyright © 1995
New Readers Press
Publishing Division of Laubach Literacy International
1320 Jamesville Avenue, Syracuse, New York 13210

Printed in the United States of America

Director of Acquisitions and Development: Christina Jagger
Photo Illustrations: Mary McConnell
Developer: Learning Unlimited, Oak Park, IL
Developmental Editor: Kathy Osmus

9 8 7 6 5

Contents

Introduction

Math skills play an increasingly vital role in today's world. Everyone needs to work confidently with numbers to solve problems on the job and in other areas of daily life.

This book and the others in the *Smart Solutions* series can help you meet your everyday math needs. Each unit is organized around four key areas that will build your competence and self-confidence:

- **Skills** pages present instruction and practice with both computation and word problems.
- **Tools** pages provide insight into how to use objects (such as rulers or calculators) or apply ideas (such as estimates or equations) to a wide variety of math situations.
- **Problem Solver** pages provide key strategies to help you become a successful problem solver.
- **Application** pages are real-life topics that require mathematics.

Key Features

Skill Preview: You can use the Skill Preview to determine what skills you already have and what you need to concentrate on.

Talk About It: At the beginning of each unit, you will have a topic to discuss with classmates. Talking about mathematics is key to building your understanding.

Key Concepts: Throughout the book, you will see this symbol ▶, which indicates key math concepts and rules.

Making Connections: Throughout each unit, you will work with topics that connect math ideas to various interest areas and to other math concepts.

Special Problems: These specially labeled problems require an in-depth exploration of math ideas. You may be asked to explain or draw or to do something else that demonstrates your math skills.

Working Together: At the end of each unit, you will work with a partner or small group to apply your math skills.

Mixed and Unit Reviews: Periodic checkups will help you see how well you understand the material and can apply what you have learned.

Posttest: At the end of the book, you will find a test that combines all of the book's topics. You can use this final review to judge how well you have mastered the book's skills and strategies.

Glossary: Use this list of terms to learn or review key math words and ideas.

Tool Kit: You can refer to these resource pages as you work through the book.

Skill Preview

This survey of math skills will help you and your teacher decide what you need to study to get the most out of this book. It will show you how much you already know and what you need to learn in the areas of whole numbers and money.

Do as much as you can of each section below. If you can't do all of the problems in a section, go ahead to the next section and do all of the problems that you can.

Part 1: Using Numbers

1. Choose the number that has three digits:

   3 33 333

2. Is 25 an even or odd number?

3. Write the next term in the pattern below.

   4, 8, 12, _____

4. Choose a symbol (<, >, or =) to compare the numbers below.

   28 _____ 15

5. Write the number *four thousand fifty-six* in figures.

6. Round 2,569 to the nearest thousand.

7. What is the unit distance from point A to point B?

   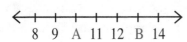

8. Independence Day is the fourth of July. On what day of the week does it fall according to the calendar?

   July

S	M	T	W	T	F	S
					1	2
3	4	5	6	7	8	9
10	11	12	13	14	15	16
17	18	19	20	21	22	23
24	25	26	27	28	29	30
31						

9. Write the time shown on the clock.

10. Match the numbers in the first column with the words and symbols in the second column.

    _____ 4¢   **a.** $.40

    _____ 40¢   **b.** sixty-four cents

    _____ $6.04   **c.** six hundred four dollars

    _____ $6.40   **d.** $.04

    _____ $64.00   **e.** six dollars and forty cents

    _____ $604   **f.** six dollars and four cents

    _____ $.64   **g.** sixty-four dollars

Part 2: Addition

Solve these addition problems.

11.
$$\begin{array}{r} 24 \\ +\ 65 \\ \hline \end{array}$$

13.
$$\begin{array}{r} 346 \\ +\ 25 \\ \hline \end{array}$$

15.
$$\begin{array}{r} 8,247 \\ +\ 3,596 \\ \hline \end{array}$$

12. 687 + 45

14. 928 + 15 + 476

16. 246,359 + 7,248

Solve the following problems.

17. In 1995 Columbus College charged $2,700 per semester for tuition. If the college increased that amount by $225 in 1996, what was the new tuition?

18. Shelly kept a log of the miles she drove delivering flowers each week. Find the total mileage for the week shown in her logbook.

Sun	Mon	Tue	Wed	Thu	Fri	Sat
0	0	18	32	50	124	87

19. To contain his livestock, Pablo fenced the rectangular pasture shown below. What is the perimeter of the field?

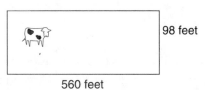

98 feet

560 feet

20. Allison must total the bill before she gives it to the customer. What is the total for this customer?

Laverne's Diner	
Item	Price
Chicken Sand.	3.95
Cola	.85
Deluxe Hamb.	4.75
Coffee	.60
Pie	2.40
Subtotal	
Tax	1.00
Total	

Part 3: Subtraction

Solve these subtraction problems.

21.
$$\begin{array}{r} 486 \\ -\ 352 \\ \hline \end{array}$$

22.
$$\begin{array}{r} 3,292 \\ -\ 1,264 \\ \hline \end{array}$$

23.
$$\begin{array}{r} 275,000 \\ -\ 48,012 \\ \hline \end{array}$$

24. 4,650
    – 349

25. 521
    – 99

26. 7,000
    – 989

Solve the following problems.

27. If Katie took 18 pictures on a roll of film that has 27 exposures, how many pictures does she have left?

29. Izzy's car has a 36-month or 36,000-mile warranty. During the past 27 months, he has driven the car 28,756 miles. How many miles remain before the warranty expires?

28. If you ride the bus to and from work each day, how much do you save by buying the round-trip ticket?

Bus Passes	
One-Way	$1.25
Round Trip	$2.25

30. Asfa used her 75¢-off coupon to buy shampoo priced at $2.98. She paid 18¢ in tax. How much change did she receive if she paid with a $5 bill?

Part 4: Multiplication

Solve these multiplication problems.

31. 31
   × 7

33. 48
   × 9

35. 720
   × 6

32. 5,000 × 90

34. 386 × 57

36. 9,108 × 43

Solve the following problems.

37. Because their director is retiring, all 25 members of the chorus chipped in $5 apiece for a gift. How much do they have to spend on the gift?

38. On Janice's map, 1 inch represents 75 miles. What is the distance between two cities that are 12 inches apart on the map?

39. Choose the best pay for a year's work from the chart below. Explain your answer.

$580 per week
$2,500 per month
$28,000 per year

40. Find the number of square feet of sod that is needed to cover with grass the backyard shown below.

98 feet

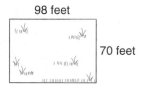

70 feet

Part 5: Division

Solve these division problems.

41. $8\overline{)728}$

43. $35\overline{)840}$

45. $15\overline{)925}$

42. $6,321 \div 7$

44. $\dfrac{2,800}{70}$

46. $6,592 \div 64$

Solve the following problems.

47. A cellular phone manufacturing plant produced 98,340 phones in 6 months. What was the average number of phones produced per month?

48. The distance from New York City to Los Angeles is 3,685 miles. How many hours would it take to drive the distance at 55 miles per hour?

49. If you can save $250 per month, how long will it take you to save $6,000?

50. Find the unit price of developing 12-exposure, 24-exposure, and 36-exposure rolls of film. Which size roll costs the least per picture to develop?

35mm
COLOR PRINTS
★5.99
24 Exposures
12 Exp.
3.99
36 Exp.
7.99

Answers on next page.

1. 333

2. odd

3. 16

4. 28 > 15

5. 4,056

6. 3,000

7. 3 units

8. Monday

9. 10:15

10. **d, a, f, e, g, c, b**

11. 89

12. 732

13. 371

14. 1,419

15. 11,843

16. 253,607

17. **$2,925**

$2,700 + $225 = $2,925

18. **311 miles**

18 + 32 + 50 + 124 + 87 = 311

19. **1,316 feet**

98 + 98 + 560 + 560 = 1,316

20. **$13.55**

$3.95 + $0.85 + $4.75 + $0.60 + $2.40 = $12.55

$12.55 + $1.00 = $13.55

21. 134

22. 2,028

23. 226,988

24. 4,301

25. 422

26. 6,011

27. **9 pictures**

27 − 18 = 9

28. **25¢ = $0.25**

$1.25 + $1.25 = $2.50

$2.50 − $2.25 = $0.25

29. **7,244 miles**

36,000 − 28,756 = 7,244

30. **$2.59**

$2.98 − $.75 = $2.23

$2.23 + $.18 = $2.41

$5.00 − $2.41 = $2.59

31. 217

32. 450,000

33. 432

34. 22,002

35. 4,320

36. 391,644

37. **$125**

$5 × 25 = $125

38. **900 miles**

12 × 75 = 900

39. **$580 per week**

$$\begin{array}{ll} \$580 & \$2,500 \\ \underline{\times\ 52} & \underline{\times\ 12} \\ 1\ 160 & 5\ 000 \\ \underline{29\ 00} & \underline{25\ 00} \\ \$30,160 & \$30,000 \qquad \$28,000 \text{ per year} \\ \text{Best} & \end{array}$$

40. **6,860 square feet**

98 × 70 = 6,860

41. 91

42. 903

43. 24

44. 40

45. 61 R10

46. 103

47. **16,390 phones per month**

98,340 ÷ 6 = 16,390

48. **67 hours**

3,685 ÷ 55 = 67

49. **24 months**

$6,000 ÷ $250 = 24

50. **36 exposures cost the least to develop at $.22 per picture.**

12 exposures: $3.99 ÷ 12 = $0.3325 ≈ $.33

24 exposures: $5.99 ÷ 24 = $0.249 ≈ $.25

36 exposures: $7.99 ÷ 36 = $0.2219 ≈ $.22

Skill Preview Diagnostic Chart

Make note of any problems that you answered incorrectly. Notice the skill area for each problem you missed. As you work through this book, be sure to focus on these skill areas.

Problem Number	Skill Area	Unit
1, 4	Comparing numbers	1
2	Even and odd numbers	1
3	Patterns	1
5	Writing numbers	1
6	Rounding numbers	1
7	Using a number line	1
8	Reading a calendar	1
9	Telling time	1
10	Reading money amounts	1
11, 12, 13, 14, 15, 16	Adding numbers	2
17, 18	Understanding word problems	2
19	Finding perimeter	2
20	Adding dollars and cents	2
21, 22, 23, 24, 25, 26, 27	Subtracting numbers	3
28, 30	Multistep	3
29	What do I have to find?	3
31, 32, 33, 34, 35, 36, 37	Multiplying by one digit	4
38	What information do I need?	4
39	Multistep	4
40	Finding area	4
41, 42, 43, 44, 45, 46, 48, 49	Dividing numbers	5
47	Finding an average	5
50	Finding unit price	5

Using Numbers

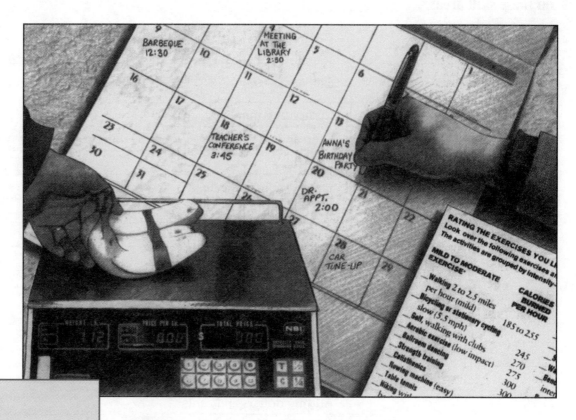

Welcome to the world of numbers! You can have fun with numbers. Turn the book upside down to read the greeting below.

$$0.7734$$

Numbers can be useful as well as fun.

Did you

- step on a scale this morning?
- buy a cup of coffee?
- notice today's temperature?
- telephone a friend?

These actions all require **number sense**—knowing what a number means and how to use it. In this unit you will increase your number sense by using numbers as you do in daily life. You will read and write numbers, talk about numbers, and work with numbers.

Everyday Numbers

Let's look at the numbers that you use every day. Fill in the survey below with as much information as you can. Write on the lines to the right of the questions.

Personal Survey

__B__ 1. What year is it now? _____

_____ 2. What is your phone number? _____

_____ 3. What is your zip code? _____

_____ 4. What is your Social Security number? _____

_____ 5. How much do you spend on a candy bar? _____

_____ 6. In what year were you born? _____

_____ 7. How many miles is your birthplace from here? _____

_____ 8. If a gas tank is empty, how many gallons of gas are in it? _____

_____ 9. What time is your favorite TV show? _____

_____ 10. What channel is it on? _____

_____ 11. What is the speed limit in front of your home? _____

_____ 12. What is the population of your town? _____

_____ 13. What time did you go to bed last night? _____

_____ 14. What are the make and year of your favorite kind of car? _____

_____ 15. How much do you think it costs? _____

_____ 16. What is your ideal weight? _____

_____ 17. How many hours do you usually sleep at night? _____

_____ 18. How much money would you like to take on a shopping spree? _____

_____ 19. How many hours a week are you going to practice your math? _____

_____ 20. What is your favorite number? _____

Talk About It

Decide how the numbers in the survey are used. Before each question, write the letter of one of the categories below to show the function of each number. Compare your answers with a partner. The first one is done for you.

A. Number that **identifies** a place or person

B. Number that tells **when**

C. Number that tells **how many**

D. Numbers that tell **how much**

E. Just a number

Success in Math

Are you ready to get involved in learning math? If the answer is yes, then this book is for you.

Many people go through life unsure of their basic math skills. They feel guilty, embarrassed, and sometimes even angry. This is often called *math anxiety*. If this has happened to you, then you are ready for a fresh start in math.

This book is not a novel. You do not just read it. You work your way through it. You ask questions. You try things with paper and pencil, calculators, and many other tools.

A. Explore your feelings about math. Mark each statement true (*T*) or false (*F*).

_____ **1.** I will never learn math.

_____ **2.** I don't like math.

_____ **3.** Everyone else seems to know what to do in math.

_____ **4.** Math has to be done quickly.

_____ **5.** I can't use my fingers to figure things out.

_____ **6.** Math books are hard to read.

_____ **7.** I don't use numbers very much.

_____ **8.** Other people make math look easy.

Did you answer *true* to any of the statements on page 14? If so, now is the time to face these negative feelings and begin the change to a positive math attitude. One of the goals of this book is to help you develop a positive attitude toward math so you can experience some of the statements in the checklist below.

B. Here is a checklist of positive statements about math. Check (✓) the ones you think will help you succeed in math.

☐ I will learn math one step at a time.

☐ I will keep a math notebook.

☐ I will use paper and pencil as I work through my math book.

☐ I will use math in real-life situations.

☐ I will enjoy working with numbers.

☐ I will take my time and relax when I work with numbers.

☐ Good math skills will make me feel good.

☐ My math will improve each day.

☐ I will practice my math skills every day.

☐ Math helps me in daily life.

☐ I will build on the math skills I already have.

☐ I have a chance to be successful in math.

As you work through this book, enjoy your success.

There are several things you can do to be a winner at math:

• Use the Answer Key in the back of the book to check your work.

• Take advantage of the Tool Kit. It has useful information and suggestions, such as addition and multiplication tables, measurement charts, and examples and definitions of many math symbols and ideas.

• Keep a math notebook and pencil at hand as you study. Use it to write down ideas and questions.

• Practice using math tools to help you solve problems.

• Continue reviewing skills you have already studied.

• Do math in your daily life to make math meaningful.

One more thing: be sure to copy numbers accurately when you work on paper. And double-check your answers when possible.

Counting and Grouping

Counting is one of the first math skills people learn. Even before we had a number system, people counted things by making notches on a stick or by piling pebbles.

Our number system is a base ten system using symbols. ***Base ten*** means there are 10 symbols: 0, 1, 2, 3, 4, 5, 6, 7, 8, and 9. Each symbol is called a **digit**, the Latin word meaning "finger."

Examples: 5 is a **one-digit number.**

73 is a **two-digit number.**

594 is a **three-digit number.**

A. How many digits are used to write each of these numbers?

1. 29 _____   **2.** 12,586 _____   **3.** 947 _____   **4.** 7 _____

The most common way to count is by *ones:* 1, 2, 3, 4, 5, . . . The three dots after a series of numbers mean the numbers continue even though they are not written. Sometimes we use other **patterns** to count. We could count by *twos:* 2, 4, 6, 8, 10, . . . Or by *fives:* 5, 10, 15, 20, 25, . . .

B. Practice counting by patterns.

5. Count by tens to 90.

6. Count by hundreds to 900.

7. Continue these counting patterns:

   a. 3, 6, 9, ___, ___, . . .   **b.** 4, 8, 12, ___, ___, . . .   **c.** 30, 24, 18, ___, ___

You may know the **tally system** of counting. The tally system can help you record data at work. For example, a salesclerk might record the sale of each T-shirt by color.

8. Notice that for every group of 5, the 5th slash crosses a group of 4. Find the total for each color by counting by fives.

T-Shirt Color Chart

White	Gray	Black	Red	Blue	Yellow
卅	卅	卅	卅	卅	III
卅	卅	IIII	II	卅	
卅	I			II	
III					
total	total	total	total	total	total

Discuss Explain how you found the totals. How else could you group the numbers?

16

Digits

A digit can be used to describe a group of ones, tens, hundreds, etc.

- A one-digit number shows how many ones: 8 means 8 ones.
- A two-digit number shows tens and ones: 34 means 3 tens 4 ones.
- A three-digit number shows hundreds, tens, and ones: 205 is 2 hundreds 0 tens 5 ones.

C. Use digits to write the numbers below.

9. 4 hundreds 3 tens 2 ones _____

10. 6 tens 7 ones _____

11. 9 hundreds 2 tens 0 ones _____

12. 2 hundreds 5 ones _____

13. 7 tens _____

14. 3 hundreds 7 tens _____

D. Answer the problems about numbers.

15. How many digits are used to write each number?

   a. 576,342 _____   b. 58 _____   c. 296 _____   d. 4,793 _____

16. Write the missing number in each counting pattern.

   a. 4, 8, 12, _____, 20, 24, . . .   b. _____, 5, 6, 7, 8, . . .   c. 20, 25, _____, 35, _____, . . .

17. In the number 784, which digit shows the number of tens?

Making Connections: Odd and Even Numbers

Counting by twos gives you a number pattern of **even numbers:** 2, 4, 6, 8, 10, 12, . . . Any number whose last digit is 0, 2, 4, 6, or 8 is an even number. Even numbers can be used to describe things in pairs.

Sometimes when you put items in pairs, there aren't enough items to work out evenly. Whole numbers that are not even are called **odd.** These numbers are 1, 3, 5, 7, 9, 11, . . .

1. Tell whether each number below is even or odd.

   a. 96 _____   b. 485 _____   c. 9 _____   d. 248 _____   e. 156,327 _____

2. If every member of the marching band can be paired with a partner, which number can describe the size of the band?

   (1) 72 members   (2) 73 members   (3) 71 members   (4) 75 members

3. **Explain** Rinji took 15 socks out of the dryer. Will any socks be left over after he pairs them? Explain how you arrived at your answer.

The Number Line

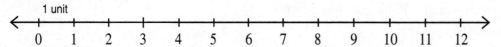

Arithmetic is the study of numbers. Geometry is the study of shapes. One of the simplest shapes in geometry is a **line.** We can use a **number line** to show the order of numbers.

1 unit
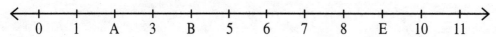

The arrows show that the line continues without end. Equal units are marked on the line. On this line, each mark represents a whole number. Notice that the numbers' values get larger as you go to the right.

Number lines can differ. Number lines do not always begin at zero. Units can be larger or smaller than 1. Look at these examples.

Understanding Number Lines

Examples:

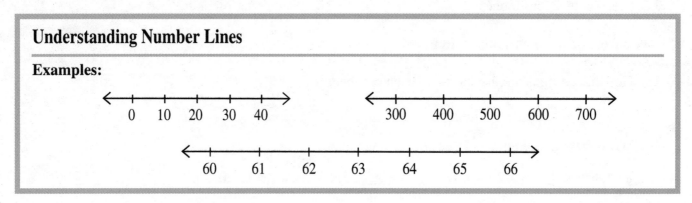

A. Answer the problems using the number line below.

1. What number is located at point A?

2. What number is located at point E?

3. How many units are there between 2 and 7 on this number line?

4. What number is 3 units to the right of 5 on the line?

5. What number is 1 unit to the left of 6 on the line?

6. How many units are there between point B and point E?

7. As you go to the left on a number line, do the numbers get larger or smaller?

The number line is a tool you can use to compare numbers. The line puts numbers in order from smallest to largest. The **symbols** you use to show the comparisons are

7 = 7
equals

7 > 2
is greater than

7 < 10
is less than

B. Compare the pairs of numbers below. Use =, >, or < .

   8. 9 _____ 6   **9.** 15 _____ 15   **10.** 3 _____ 12   **11.** 4 _____ 2   **12.** 100 _____ 88

C. Write each sentence below using numbers and symbols.

   13. Three is greater than one.   **14.** Forty-eight is less than forty-nine.   **15.** Nine equals nine.

Making Connections: Measuring Tools

Number lines are used on many common measuring tools. One example is a **yardstick.**

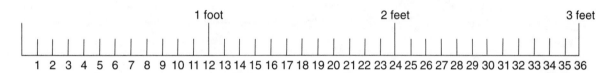

Answer these questions about a yardstick.

   1. The small units on a yardstick are inches. How many inches are shown on the yardstick?

   2. The large units on a yardstick are feet. How many feet are marked on the yardstick?

   3. Fill in the blanks. Find your answers by looking at the yardstick.

      a. 1 foot = _____ inches   **b.** 24 inches = _____ feet   **c.** 3 feet = _____ inches

Thermometers and **scales** are also number lines found in daily life.

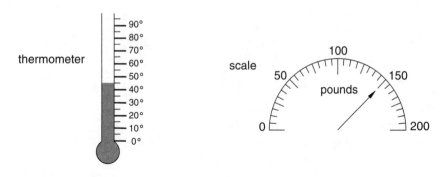

   4. The units on the thermometer are degrees. What is the temperature shown on the thermometer? _____ degrees

   5. The units on the scale are pounds. What weight is shown on the scale? _____ pounds

   6. What other number lines do you see in your daily life?

For another look at measuring tools, turn to page 207.

Place Value

When you read, write, and say numbers, you use **place value.** The value of a number depends on the placement of its digits. The place value of a three-digit number is shown below. The number **428** shows how to use the place value chart.

428

Places	hundreds	tens	ones
Digits	4	2	8
Values	400	20	8

four hundred twenty-eight

Examples: Rearranging the digits changes the value of the number.

284

Places	hundreds	tens	ones
Digits	2	8	4
Values	200	80	4

two hundred eighty-four

842

hundreds	tens	ones
8	4	2
800	40	2

eight hundred forty-two

A. Use your understanding of place value to solve these problems.

1. Fill in the missing information in the charts. Write the numbers in words below the charts.

Places	hundreds	tens	ones
Digits	7	9	5
Values			

hundreds	tens	ones
300	50	0

_____   _____

Places			ones
Digits	8	5	
Values	800	50	6

_____

2. Match the words in the second column with the numbers in the first column.

_____ 573   **a.** five hundred thirty-seven

_____ 753   **b.** seven hundred fifty-three

_____ 735   **c.** five hundred seventy-three

_____ 537   **d.** seven hundred thirty-five

B. For more practice, solve the problems below.

3. In the number 386, which digit is in the tens place?
   What is its value?

4. Write the number one hundred seventy-six in figures.

5. Match the words in the second column with the numbers in the first column.

   _____ 940   a. forty-nine
   _____ 409   b. nine hundred four
   _____ 904   c. four hundred ninety
   _____ 49   d. nine hundred forty
   _____ 490   e. four hundred nine

Naming Larger Numbers

Larger numbers use a larger place value chart like the one below. Commas help you read larger numbers. Starting from the right, use commas to separate every three digits into place value groups.

Example: 7040348 is written as **7,040,348.**

Places	millions	thousands	hundreds	tens	ones
Digits	7	040	3	4	8
Values	7 million	40 thousand	300	40	8

seven million forty thousand three hundred forty-eight

C. Choose the number that matches the given value.

6. Which number is three million four hundred twenty thousand fifteen?
   (1) 342,015   **(2)** 34,202,115   **(3)** 3,420,015

7. Which number is twelve thousand nine hundred two?
   (1) 12,920   **(2)** 12,902   **(3)** 120,902

D. To compare and order larger numbers, remember that the number with the greatest place value is the largest.

8. Compare the pairs of numbers below. Put =, <, or > in the box.
   a. 453 ▮ 456   **b.** 72,302 ▮ 72,032   **c.** 108,343 ▮ 180,342

9. Order these numbers from *smallest to largest.*
   a. 28   208   82   2,008
   b. 2,345   2,354   2,435   2,453
   c. 765   576   6,057   60,507

> **For another look**
> at symbols, turn to
> page 205.

Rounding

Numbers are often **rounded** to approximate values when exact numbers are not necessary. Below are just two examples of rounded numbers. Why do you think the numbers are called rounded numbers?

- The distance from the Earth to the Moon is approximately 240,000 miles.

- You can receive close to 50 channels on your cable TV.

You can use a number line to help you round numbers. Numbers can be rounded up or down to the nearest ten, nearest hundred, nearest thousand, and so on. A place value that holds a 5, such as 15 or 150, rounds up to a higher number.

Using Number Lines

Example: Round the number 24 to the nearest ten. On the number line, it is located between 20 and 30 as you count by tens. This can be shown by using the symbol <, meaning *is less than*.

$$20 < 24 < 30$$

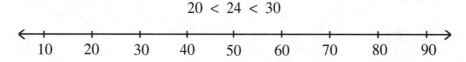

To round 24 to the nearest ten, **round down to 20** because 24 is closer to 20 than to 30.

A. Using the number line above, round the following numbers to the nearest ten. The first one is started for you.

	Exact	Between	Round Up or Down	Rounded Number
1.	19	10 < 19 < 20	_____	_____
2.	48	____ < 48 < ____	_____	_____
3.	71	____ < 71 < ____	_____	_____
4.	82	____ < 82 < ____	_____	_____

B. Use this number line to round these numbers to the nearest hundred.

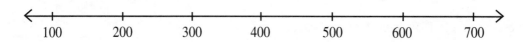

5. a. 125 rounds to _____   **b.** 386 rounds to _____   **c.** 561 rounds to _____

Rounding a Number

Examples: You can round a number to any place value. Round 1,238 and 1,274 to the nearest *hundred*.

Step 1
Find the digit in the place being rounded.

1,2̲38   1,2̲74

Step 2
Look at the next digit to the right. If the digit is less than 5, *round down.*

1,2③8   1,2⑦4

3 < 5   7 > 5   If the digit is 5 or more, *round up.*

1,200   1,300

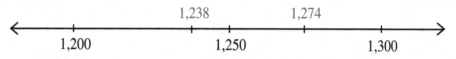

1,238 is closer to 1,200.   **1,274 is closer to 1,300.**

C. Practice rounding numbers.

6. Round the numbers below to the nearest ten.
   a. 48   **b.** 178   **c.** 93   **d.** 672   **e.** 325

7. Round the numbers below to the nearest hundred.
   a. 4,839   **b.** 12,576   **c.** 752   **d.** 808   **e.** 555

8. Round the numbers below to the nearest thousand.
   a. 2,510   **b.** 28,391   **c.** 7,145   **d.** 13,383   **e.** 1,789

Making Connections: Real-Life Estimation

In daily life you use rounding to estimate an answer. Go shopping below using rounded prices. Estimate how much money you will need to make each purchase.

   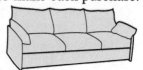

1. $22,389 is close to _____.
         estimate

2. $795 is close to _____.
         estimate

Does the Answer Make Sense?

Leon is 18 feet tall.

The giraffe at the zoo is 18 feet tall.

Jean's grandfather is 18 years old.

Jean's teenage son is 18 years old.

Take 18 teaspoons of medicine and call the doctor
in the morning.

Take 2 teaspoons of medicine 3 times a day for 18 days.

For which of the situations does the answer 18 make sense?

It *does* make sense

- for a giraffe to be 18 feet tall

- for a teenager to be 18 years old

- to take 2 teaspoons of medicine 3 times
  a day for 18 days

It *does not* make sense

- for someone to be 18 feet tall

- for a grandfather to be 18 years old

- to take 18 teaspoons of medicine

When solving math problems, you have to look at each answer you get and ask
yourself, "Does the answer make sense?" If not, this is a signal to try solving the
problem again. If the answer is sensible, and you have carefully worked the math,
then your answer is probably correct.

A. Select the most sensible number to describe each situation.

1. The average adult has _____ quarts of blood.
   (1) 6   **(2)** 600   **(3)** 60,000

2. The temperature on a warm fall afternoon in Illinois was _____.
   (1) 97°F   **(2)** 72°F   **(3)** 27°F

3. The population of the United States is about _____ people.
   (1) 2,500   **(2)** 25,000   **(3)** 250,000,000

4. The speed limit on the expressway is _____ miles per hour.
   (1) 55   **(2)** 505   **(3)** 555

5. My grandmother lived to be _____ years old.
   (1) 8   **(2)** 80   **(3)** 800

B. When shopping for a new or used car, people usually estimate how much money they are willing to spend. Round each car's sticker price to the nearest thousand and match it to a customer. Be sure there is a car for each customer.

Claudia's Car Dealership

Family Sedan costs $14,285.

Coupe costs $16,394.

Minivan costs $22,941.

Super Sport costs $27,768.

Mazher has $18,000 to spend.

Ben has $15,000 to spend.

Miguel has $25,000 to spend.

Betty has $30,000 to spend.

C. Look at each answer again. Explain why each answer makes sense.

D. There are times when an exact answer is very important. In each problem below, decide if you would need an exact answer or an estimate. Explain your answers.

6. What time does the plane leave for Chicago?

    (1) estimate: about 3 o'clock   **(2)** exact: 2:48 P.M.

7. What is the inseam on your slacks?

    (1) estimate: about 30 inches   **(2)** exact: 32 inches

8. When talking to a friend, you ask, "How many people attended the church service?"

    (1) estimate: around 200 people   **(2)** exact: 195 people

9. How tall is the pine tree next to the house?

    (1) estimate: approximately 30 feet   **(2)** exact: 28 feet

10. To figure the total for your paycheck, how many hours were on your timecard at work last week?

    (1) estimate: about 40 hours   **(2)** exact: 43 hours

Mixed Review

A. **Use your knowledge of numbers to answer the questions.**

1. Think of the numbers in your life. Choose four that are important to you and explain their importance.

   a. _____ Why? _____

   b. _____ Why? _____

   c. _____ Why? _____

   d. _____ Why? _____

2. What would you say to encourage somebody who has math anxiety?

3. How many digits are used to write each number below?
   a. 76 **b.** 3,193 **c.** 2

4. Fill in the missing numbers in the counting patterns below.
   a. 16, 17, _____ , 19, 20, _____ , 22, 23, . . .
   b. 3, 6, _____ , 12, _____ , 18, 21, . . .

5. Are the numbers below even or odd?
   a. 78 **b.** 203 **c.** 1,042 **d.** 15

6. There are 76 people in the dance club. Can everyone have a partner? How do you know?

7. Use this number line to answer the questions below.

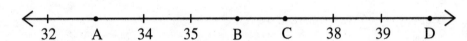

   a. What number is located at point C?
   b. What is the distance between point A and point D?
   c. What number is 2 units to the left of point B?

B. **Compare the pairs of numbers below using =, >, or <.**

8. **a.** 47 _____ 39 **b.** 9 _____ 56 **c.** 46 _____ 46

9. **a.** 286 _____ 280 **b.** 3,210 _____ 3,201 **c.** 704 _____ 740

C. Solve the problems involving number values.

10. Match the words in the second column with the numbers in the first column.

_____ 59,300   **a.** five thousand ninety-three

_____ 5,903   **b.** fifty-nine thousand three hundred

_____ 5,093   **c.** five thousand nine hundred three

_____ 5,930   **d.** five thousand nine hundred thirty

11. Using each of the digits 2, 8, 3, and 7 only once:
   a. write the *largest* possible number
   b. write the *smallest* possible number

12. a. Round 5,262 to the nearest thousand.
   b. Round 4,941 to the nearest thousand.

13. List four numbers that would round to 100.

D. Fill in the blanks with approximate values. If you do not know the answer, estimate with any number that makes sense. Discuss your answers with a partner.

14. This book has about _____ pages. It is about _____ inches wide and _____ inches long.

15. I travel _____ miles to get to school. I would travel _____ miles to get here if necessary.

16. The population of my town is _____. I know approximately _____ people in the town. There are _____ stores in this town. The tallest building is _____ stories tall.

17. The most expensive car that I like would cost about _____ dollars. It would cost approximately _____ dollars to fill the gas tank. The yearly car insurance would probably be _____ dollars.

18. Look at the numbers that you have used. Which numbers are round numbers? How do you know? Do your answers make sense?

Calendars

Calendars help us organize our lives. A **calendar** can show the past, the present, and the future. The units of a calendar are **year, month, week,** and **day.**

- A year is the time the Earth takes to revolve around the Sun once.

- A day is how long the Earth takes to spin around once on its axis.

- One year is a little more than 365 days. Every 4 years the extra time is added together in a **leap year,** which has 366 days. Recent leap years have been 1988 and 1992. Is this year a leap year?

A calendar organizes the days into weeks and months. One week is 7 days. A month may be 28, 29, 30, or 31 days. Look at the calendar below for the month of November in 1999.

| \multicolumn{7}{c}{November 1999} |
S	M	T	W	Th	F	S
	1	2	3	4	5	6
7	8	9	10	11	12	13
14	15	16	17	18	19	20
21	22	23	24	25	26	27
28	29	30				

The days of the week are listed across the top. This month begins on Monday. It lasts 30 days. It ends on Tuesday. The next month will begin on Wednesday. What day is November 26?

A year has 12 months. Memorize the months and the number of days shown in the chart. Abbreviations or numbers are often used to write the months.

1	January (Jan.)	31 days		7	July	31 days
2	February (Feb.)	28 days		8	August (Aug.)	31 days
		(29 in leap year)		9	September (Sept.)	30 days
3	March (Mar.)	31 days		10	October (Oct.)	31 days
4	April (Apr.)	30 days		11	November (Nov.)	30 days
5	May	31 days		12	December (Dec.)	31 days
6	June	30 days				

A. Answer the problems about calendars and dates.

1. Which month is the fourth month?

2. Which month has the extra day for leap year?

3. How many days are in the sixth month?

4. Most of the months are how many days long?

Writing Dates

A written date can name the month, day, and year.

Example 1: November 17, 1999 (Put a comma after the day.)
          month   day year

Dates can also be abbreviated.

Example 2: Nov. 17, 1999 *or* 11/17/99

When you fill out a form, it may ask you to write the date using only numbers, as shown in the last example above. Write today's date as many different ways as you can think of.

B. Use the 1999 calendar below to answer these problems.

January	February	March	April	May	June
S M T W Th F S	S M T W Th F S	S M T W Th F S	S M T W Th F S	S M T W Th F S	S M T W Th F S
1 2	1 2 3 4 5 6	1 2 3 4 5 6	1 2 3	1	1 2 3 4 5
3 4 5 6 7 8 9	7 8 9 10 11 12 13	7 8 9 10 11 12 13	4 5 6 7 8 9 10	2 3 4 5 6 7 8	6 7 8 9 10 11 12
10 11 12 13 14 15 16	14 15 16 17 18 19 20	14 15 16 17 18 19 20	11 12 13 14 15 16 17	9 10 11 12 13 14 15	13 14 15 16 17 18 19
17 18 19 20 21 22 23	21 22 23 24 25 26 27	21 22 23 24 25 26 27	18 19 20 21 22 23 24	16 17 18 19 20 21 22	20 21 22 23 24 25 26
24 25 26 27 28 29 30	28	28 29 30 31	25 26 27 28 29 30	23 24 25 26 27 28 29	27 28 29 30
31				30 31	

July	August	September	October	November	December
S M T W Th F S	S M T W Th F S	S M T W Th F S	S M T W Th F S	S M T W Th F S	S M T W Th F S
1 2 3	1 2 3 4 5 6 7	1 2 3 4	1 2	1 2 3 4 5 6	1 2 3 4
4 5 6 7 8 9 10	8 9 10 11 12 13 14	5 6 7 8 9 10 11	3 4 5 6 7 8 9	7 8 9 10 11 12 13	5 6 7 8 9 10 11
11 12 13 14 15 16 17	15 16 17 18 19 20 21	12 13 14 15 16 17 18	10 11 12 13 14 15 16	14 15 16 17 18 19 20	12 13 14 15 16 17 18
18 19 20 21 22 23 24	22 23 24 25 26 27 28	19 20 21 22 23 24 25	17 18 19 20 21 22 23	21 22 23 24 25 26 27	19 20 21 22 23 24 25
25 26 27 28 29 30 31	29 30 31	26 27 28 29 30	24 25 26 27 28 29 30	28 29 30	26 27 28 29 30 31
			31		

5. What day of the week is 11/17/99?

6. On what day of the week does August start?

7. What is the date of the third Saturday in April?

8. How many months are there from:

   a. March 10 to June 10   **b.** September 5 to December 5

9. How many days are there from:

   a. May 3 to May 17   **b.** October 5 to October 28

10. If you leave on June 12 for a 2-week vacation, when will you return?

11. If you get paid every 2 weeks and the first pay date is August 6, when is the next payday?

12. On what day of the week will January 1, 2000, be?

Using Dollars and Cents

| $20 | $10 | $5 | $1 | penny | nickel | dime | quarter | half-dollar |

The money above shows the most common U.S. coins and bills.

$ is the symbol for dollars. For example, one dollar is $1 and twenty dollars is $20. An amount like $72 can be made using many different combinations of bills. You can use counting to select the bills you need.

Counting Dollars

Example 1: $72

bills:	$10	$10	$10	$10	$10	$10	$10	$1	$1
count:	10	20	30	40	50	60	70	71	72 = **$72**

Example 2: $72

bills:	$20	$20	$20	$10	$1	$1
count:	20	40	60	70	71	72 = **$72**

A. Practice your skills with money on the problems below.

1. Use counting to select the bills for the amounts shown. Find as many combinations of bills as possible. Record and share your results.

   a. $18

   bills:

   count:

   b. $90

   bills:

   count:

2. How much money is shown in the pictures below?

a. _____

b. _____

An amount less than one dollar is called *cents*. Cents can be written
- using the cents symbol: 39¢ or 5¢
- using a dollar sign and decimal point: $.39 or $.05

Different combinations of coins can be exchanged for the same amount. For example:

$1 = 100 pennies $1 = 20 nickels $1 = 10 dimes $1 = 4 quarters

Dollars and Cents

Dollars and cents can be combined using a dollar sign and a **decimal point.** As shown in the examples, read the decimal point by saying *and*.

Reading Dollars and Cents

Examples: $5.24 $16.05* $42.60* $85.00*

five dollars *and*   sixteen dollars *and*   forty-two dollars *and*   eighty-five dollars

twenty-four cents   five cents   sixty cents

---

*What is the purpose of the zeros? What would happen if they were not written?

B. Use dollar signs and decimal points to write the totals below.

3. **a.** $10, $5, $1, 25¢, 25¢, 1¢   **b.** $20, $20, $20, $5, 25¢, 10¢, 5¢, 1¢

4. Match the equal amounts in the first and second columns.

  _____   **a.** $4 $.04
  _____   **b.** 4¢ $40.00
  _____   **c.** 40¢ $4.00
  _____   **d.** $40 $.40

5. Underline the amount that is different from the others.

  a. seven cents   7¢ $.07 $.70

  b. three dollars   3¢ $3.00 $3

  c. eleven dollars and three cents $11.30 $11.03

6. Order the money from *smallest to largest.*

  a. $8 $.08   80¢ $80.00

  b. $4.82 $4.08 $48.20 $4.20

C. Use your skills with coins and bills for shopping. Different money combinations are possible.

7. If you didn't have the exact amount, what bills and coins could you use to pay for the items below?

  a. apples for 48¢   **b.** shoes for $29.35
      bills: bills:
      coins: coins:

8. The cash register says you get back $3.95. What coins and bills could you receive in change?

9. **Multiple Solutions** If the vending machine says Exact Change Only (No Pennies), what combination of coins do you need to buy a can of juice for 90¢? Can you use other coins?

Paying with Cash

When you pay for a purchase, you often give the cashier more money than the cost of the item. The extra money the cashier returns to you is the **change.** You can use counting to figure the change.

Figuring Change

Example 1: If you pay $1 for an item that costs 68¢, the change is figured as shown.

coins:		1¢	1¢	5¢	25¢
count:	68	69	70	75	$1.00

Example 2: You pay $25 for an item that costs $22.30. Here is the change:

coins and bills:		10¢	10¢	25¢	25¢	$1	$1
count:	22.30	22.40	22.50	22.75	23	24	$25

A. Use counting to figure the change you should receive for the purchases shown. The first item is figured for you.

1.
Item	Cost	Money Paid	Coins	Bills
a. notepad	78¢	$1.00	1¢ 1¢ 10¢ 10¢	none
b. lemons	34¢	$.50		
c. video	$4.95	$10.00		
d. suit	$56.12	$70.00		

You can also use counting to figure how much more money you need. For example, you have $56 and want to buy a lamp for $75. How much more money do you need?

bills:		$1	$1	$1	$1	$5	$10 = **$19**
count:	56	57	58	59	60	65	75

B. Use counting to figure how much more money you need to make the purchases shown below.

2.

VCR

garden tools

tires

a. Cost: $179
You have: $125
You need: _____

b. Cost: $36.25
You have: $22.50
You need: _____

c. Cost: $175.60
You have: $134.75
You need: _____

Estimating Money Amounts

You often estimate money amounts. To round to the *nearest $1*

- *round down* to the original dollar amount if you have *less than 50¢*
- *round up* to the next dollar amount if you have *50¢ or more*

Rounding to the Nearest $1

Examples: Round to the nearest $1.

$5.38 *rounds down* to **$5** because 38¢ is *less than 50¢.* $5.83 *rounds up* to **$6** because 83¢ is *more than 50¢.*

C. Round each amount to the nearest $1.

3. a. $7.25   **b.** $56.05   **c.** $4.98   **d.** $16.50

When rounding to the *nearest $10*

- *round down* if there is *less than $5* in the ones place
- *round up* if there is *$5 or more* in the ones place

When rounding to the *nearest $100*

- *round down* if there is *less than $50* in the tens place
- *round up* if there is *$50 or more* in the tens place

Rounding to the Nearest $10 or $100

Example 1: Round $79 to the nearest $10.
$79 *rounds up* to **$80** because 9 is *greater than* 5.

Example 2: Round $432 to the nearest $100.
$432 *rounds down* to **$400** because 3 is *less than* 5.

D. Round each amount to the nearest $10 or $100 as indicated.

4. $86 to the nearest $10

5. $675 to the nearest $100

6. $125 to the nearest $10

7. $249 to the nearest $100

Making Connections: Shopping

Let's go shopping in a jewelry catalog. Estimate how much money you'll pay for each item at right. Round the price to the nearest $10 or $100.

Super Buy!
$34.99
14K
Locket & Earrings Set

½ Carat
Diamond Earrings
$379
Our *Lowest*
Price Ever!

Blue Topaz, Garnet &
Amethyst Rings
$89

1. Item: _____ Actual Price: _____ Rounded Price: _____

2. Item: _____ Actual Price: _____ Rounded Price: _____

3. Item: _____ Actual Price: _____ Rounded Price: _____

Checks and Money Orders

Do you have bills to pay? If so, what are they and how do you pay them? Checks and money orders can be used to make payments. You may receive a paycheck, write a check on your own bank account, or buy a money order at a currency exchange or bank to pay a bill.

Look below to see the six important parts of a check.

1. date
2. person or business you are paying
3. amount in words (Write the number of cents over 100 as shown.)
4. amount in numbers
5. your signature
6. reason for check

```
Corey Phillips                                          208
579 Stonehaven                                     31-20/213
Hometown, IL  60007              May 12  19 94          ①

②  Pay to
   the order of  Margaret Lytle              $ 28.50      ④
③  Twenty-eight and 50/100 ~~~~~~~~~~~ Dollars

   American Bank
   245 Lake Street
   Hometown, IL  60007
⑥  Memo  shower gift              Corey Phillips          ⑤

   ⑈ 2719748321 ⑈    0000473210 ⑈    0208
```

A. Use the check above to answer the following problems.

1. What is the amount of the check?

2. Who will cash the check?

3. Who wrote the check?

4. What is the date of the check?

B. Write a $187.50 check to The Tire Clinic. Use today's date and your own signature.

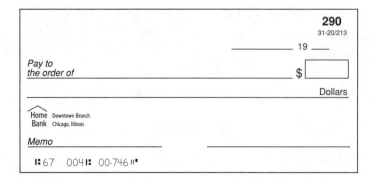

There are five important parts of a money order. Notice that your name is at the top.
An official of the bank or currency exchange signs the money order.

① date
② your name
③ person or company you are paying
④ amount in words
⑤ amount in numbers

```
Your Currency Exchange                                   2-425
1600 W. Grand Ave.                                        710
Hometown, IL  60007                  No. A 171607
338-2055                          December 10  19 95      ①

②  Remitter  Maria Flores
                                      Dollars  Cents
③  Pay to                                              ⑤
   the order of  Three Lakes Lumber Company    182  00

④  One hundred eighty-two                      Dollars

                              A.M. Amirth
State regulated               Grand Check Cashiers, Inc.
```

C. Use the money order above to answer the following problems.

5. What is the amount of the money order?

6. Who will receive payment from the money order?

7. Who is paying the bill?

8. Whose signature is on the money order?

D. Fill in the money order below to pay the bill from Dr. Omori. Use today's date.

```
Lydia Omori, MD
1300 Central Road
Hometown, IL  60007
```

Exam	$38.00
X-ray	102.00
Bandage/Supls	73.00
Total	$213.00

```
Your Currency Exchange                                   2-425
1600 W. Grand Ave.                                        710
Hometown, IL  60007                  No. A 171608
338-2055
                                     _____ 19____
Remitter _____
                                      Dollars  Cents
Pay to
the order of _____

_____  Dollars

                              A.M. Amirth
State regulated               Grand Check Cashiers, Inc.
```

What Time Is It?

Sometimes you may feel that clocks rule your life. You plan your days and nights by the time on the clock. Two kinds of clocks are most common.

The Digital Clock

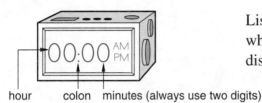

hour   colon   minutes (always use two digits)

List different objects or places where you can find digital clock displays.

Examples: 10:03 P.M.   3 minutes after 10 o'clock in the evening

10:30 A.M.   30 minutes after 10 o'clock in the morning

The Analog Clock

To read this kind of clock, you need to know these facts:

60 minutes = 1 hour

24 hours = 1 day

The clock goes through two 12-hour cycles in one day. The hour hand (the shorter hand) points to the hour. During a 60-minute period, the shorter hand moves slowly to the next hour.

Notice the 60 small units around the face of the clock. Each unit represents 1 minute. Each large hour unit marks off 5 minutes. Counting by fives can help you move quickly around the face of the clock. The minute hand (the longer hand) points to the minutes.

A. Write the digital time below each analog clock.

1. a.

b.

c.

Timely Tips

- Because there are 60 minutes in an hour, 30 minutes is half of an hour. You can say that 4:30 is *half past* four.
- Likewise, 15 minutes is a quarter of an hour. 3:15 is a quarter *after* or *past* 3 and 2:45 is a quarter *to* 3.
- A.M. is the morning.
- P.M. is the afternoon and evening.

Digital clocks show time as follows:
- Noon is 12 o'clock P.M. in the midday.
- Midnight is 12 o'clock A.M. It begins the morning hours.

B. Fill in the time on the digital clocks. Be sure to show whether the time is A.M. or P.M.

2. a. Time for breakfast.　　**b.** Let's go out for dinner.　　**c.** I have a doctor appointment this afternoon.

　　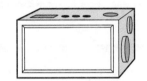

3. a. Happy New Year!　　**b.** Time for my coffee break.　　**c.** I can't be late for night class.

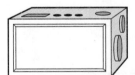　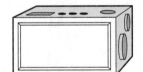　

C. Match the time in the first column to the correct words in the second column.

4. _____　10:15　　　　**a.** noon

_____　9:45　　　　**b.** quarter to ten

_____　10:30　　　　**c.** ten in the morning

_____　10:00 A.M.　　**d.** midnight

_____　10:00 P.M.　　**e.** quarter after ten

_____　12:00 P.M.　　**f.** half past ten

_____　12:00 A.M.　　**g.** ten in the evening

Tables and Schedules

Tables and schedules are charts that organize facts and numbers in **rows** across and **columns** up and down.

Table: Yearly Cost of World's Lowest Term Life Insurance

Male Age	$100,000	$250,000	$500,000	$1,000,000
35	$124	$195	$335	$610
40	$134	$210	$360	$660
45	$147	**$245**	$430	$800
50	$193	$335	$610	$1160
55	$245	$475	$890	$1720
60	$382	$738	$1415	$2770

Reading a Table

Example: In this table, a male, age 45 (row 3 across), will pay $245 per year for a $250,000 (column 3 down) life insurance policy.

A. Use the table above to answer the problems below.

1. How much will a male, age 60, pay for $250,000 in insurance?

2. At what age does a male pay the least amount for a $1,000,000 policy?

Schedule: Aerobics Classes

Day	8:00 – 8:45 AM	9:00 – 9:45 AM	10:00 – 10:45 AM	Noon – 1:00 PM	1:00 – 1:45 PM	2:00 – 2:45 PM
Monday	Wake Up	Cardio	Dance	Lunch Bunch	Work Out	Dance
Tuesday		Low	Step		Work Out	
Wednesday	Wake Up	Cardio	Dance	Lunch Bunch		Sweat
Thursday		Low			Work Out	
Friday	Wake Up	Cardio	Step	Lunch Bunch	Work Out	Dance

The Friday aerobics class from 1:00 to 1:45 P.M. is Work Out.

B. Use the schedule above to answer the problems.

3. What class is on Tuesday from 10:00 to 10:45 A.M.?

4. What time and days are Cardio classes held?

5. What are the days and times for the class that you would like to take?

C. Refer to the suburban bus schedule below to answer problems 6–8.

① RIVER ROAD CTA STATION	② NORTHWEST POINT	③ ARLINGTON HGTS & DEVON AVE.	④ THE CHANCELLORY (WYNDHAM HOTEL)
5:55 AM	6:10 AM	6:20 AM	6:29 AM
6:35	6:53	7:05	7:14
6:50	7:08	7:20	7:29
7:10	7:28	7:40	7:49
7:35	7:53	8:05	8:14
8:05	8:23	8:35	8:44
8:35	8:53	9:05	9:14
——	——	4:00 PM	4:09 PM
4:05 PM	4:25 PM	4:40	4:49
4:30	4:50	5:05	5:14
5:05	5:25	5:40	——
5:35	5:55	6:10	——
6:05	6:25	6:40	——

6. If you get to the Northwest Point bus stop at 8:00 A.M., at what time will the next bus leave?

7. Sarah missed the River Road CTA Station bus at half past four. What time does the next bus leave?

8. If you arrive at the Chancellory bus stop at 5:20 P.M., what time will you catch the next bus?

Making Connections: Daily Nutrition

This nutrition table explains the basic diet needs of people of different ages. The table is based on the U.S. Department of Agriculture's food guidelines. The serving size depends upon the type of food.

1. How many servings of vegetables should an adult woman have per day?

2. According to the table, which people should eat 6 ounces of meat per day?

3. Which groups should consume the most calories per day?

4. How many servings of fruits and vegetables should school-age children eat?

5. **Investigate** Find your place on the chart. List the foods you ate yesterday. Do you think you followed the guidelines?

Number of daily servings, based on age	Bread	Fruits	Vegetables	Meats	Milk Products	Daily Calories
School-Age Children	9	3	4	6 ounces	2	2,200
Teenage Girls	9	3	4	6 ounces	3	2,200
Teenage Boys	11	4	5	7 ounces	3	2,800
Adult Women	6	2	3	5 ounces	2	1,600 - 2,200
Adult Men	9	3	4	6 ounces	2	2,200 - 2,800
Seniors	6	2	3	5 ounces	men: 2 women: 3	1,600

Unit 1 Review

A. Use the digits 4, 3, 2, and 6 in the following problems.

1. Write the largest possible number using each of the four digits once.

2. Write the smallest number using each of the four digits once.

3. Write an odd number using each of the four digits once.

4. **a.** In the number 4,326, which digit is in the tens place?

   b. the thousands place?

5. Compare the following numbers using =, >, <.

   a. 32 _____ 23   **b.** 432 _____ 634   **c.** 2,436 _____ 2,346

6. Write the number four million three hundred six thousand. _____

7. Round each number to the *nearest hundred* and explain your reasoning.

   a. 3,426 _____   Reason: _____

   b. 3,462 _____   Reason: _____

B. Answer the problems about calendars and time.

8. Using the calendar for March 1999 below, answer the following questions.

   a. St. Patrick's Day is celebrated on March 17. What day of the week does that fall on?

   b. How many Sundays are there in March? Is that an even or odd number?

   c. On what day of the week will April Fool's Day (April 1) fall?

   d. If you play cards on the second and fourth Thursday of every month, write the dates on which you will play in March.

March

S	M	T	W	T	F	S
	1	2	3	4	5	6
7	8	9	10	11	12	13
14	15	16	17	18	19	20
21	22	23	24	25	26	27
28	29	30	31			

C. Practice your skills on these problems.

9. Answer the following questions. If you have coins totaling 76¢:

   a. What coin must you have?
   How do you know?

   b. What is the smallest number of coins you can have?
   What are the coins?
   Explain your answer.

   c. What is the largest number of coins you can have?
   Explain your answer.

10. Write the time shown on each clock below. Be sure to indicate A.M. or P.M.

a. Time for the baby's morning nap.

b. What time is the evening movie?

11. Match the numbers in the first column with the words and symbols in the second column.

_____	7¢	**a.** $.70
_____	70¢	**b.** eight dollars and three cents
_____	$83.00	**c.** $.07
_____	$8.03	**d.** eight dollars and thirty cents
_____	$.83	**e.** eighty-three dollars
_____	$8.30	**f.** eighty-three cents
_____	$803	**g.** eight hundred three dollars

12. Use counting and rounding to answer the problems below.

a. You pay $15 for a compact disc that costs $12.95. How much change should you get back? Name the bills and coins.

b. You want to buy a new sweater for $36.46. If you have $18, how much more money do you need?

13. Estimate how much you might spend on the items below.

a. 20 pounds of dog food for $1.25 per pound

b. 52-inch color projection TV for $1,875 and VCR for $235

D. Use the table to answer these problems.

This table shows the number of calories people of different weights burn in 30 minutes of continuous work.

14. a. How many calories would a 130-pound person scrubbing floors burn in 30 minutes?

b. **Estimate** Use rounded numbers to pick a combination of activities a 190-pound person could do to burn approximately 1,000 calories in a day.

Activity	Weight in Pounds / Calories Burned				
	110	130	150	170	190
Typing, electric	42	48	54	63	69
Hand sewing	48	57	66	75	84
Machine sewing	69	81	93	105	117
Cooking	69	81	93	105	117
Stocking shelves	81	96	111	126	138
Mopping	93	111	126	144	159
Painting, outside	117	129	156	177	198
Scrubbing floors	165	192	222	252	282
Mowing grass	168	198	228	258	288

Working Together

With a partner, discuss which activities listed in the table above you would choose to do. Do those activities burn the most calories? What factors were important in choosing an activity? For how long would you choose to do each activity?

Write a word problem based on the chart.

Addition

Skills

Addition facts

Adding and regrouping

Estimating

Working with dollars and cents

Tools

Addition table

Number line

Calculator

Problem Solvers

Addition strategies

Addition equations

Understanding word problems

Applications

Finding perimeter

Estimating cost

A ddition is a skill we use every day. Addition lets us combine numbers. You already know that the plus sign + is the symbol for addition.

Let's explore how you can use addition in your life.
Add the points to find the score in this football game.

Bears Score		Lions Score	
touchdown	6 points	field goal	3 points
extra point	1 point	safety	2 points
field goal	3 points	field goal	3 points

Who won? (The Bears won 10 to 8.)

In this unit you will develop your addition skills. With practice, you will increase your success in problem solving.

When Do I Add?

The use of addition is so common that people are often unaware that they are performing this operation. Whether you add sugar and cream to your coffee or build an addition to your house, you are using addition.

We add things together every day. Describe a situation in each of the settings below that involves *addition*. You may use some of the words that can mean addition, such as *sum, total, and, more, plus, increase,* and *in all.*

Examples: I always increase the amount of garlic in a recipe.
I took in a stray cat so I have 3 cats in all.

1. At home: _____

   _____

2. At school: _____

   _____

3. At the store: _____

   _____

4. At work: _____

   _____

Review the situations you described above. Do the things that you added together have anything in common? In the example, for instance, 1 cat was added to 2 cats for a total of 3 cats.

Think about what kinds of things you can combine. Can you add apples to oranges? What's the result?

This brings up an interesting point. *You can only combine things that are alike.* For instance, you can add 3 apples to 5 apples to get a total of 8 apples. However, if you want to combine 3 apples and 5 oranges, you need to find a common name that is suitable for both. You might say that 3 pieces of *fruit* plus 5 pieces of *fruit* equals 8 pieces of *fruit*.

Talk About It

Think about possible common names you could use to combine the following:

- 6 children and 4 adults

  _____

- 2 trucks and 4 cars

  _____

- 1 rose and 3 carnations

  _____

Compare and discuss your answers with a partner.

Building an Addition Table

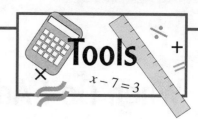

Building an addition table helps you explore addition. Begin by using counting to set up the table. Follow these steps:

1. List all the digits 0 to 9 across the top row.

2. List all the digits 0 to 9 down the first column.

3. Copy the top row into the second row. Why?

4. Copy the first column into the second column. Why?

A. Fill in the chart below by counting by ones.

Example: Part of the column under 3 is filled in as an example. Notice the order: 3, 4, 5, 6, 7, . . . Another example is the row across from 6: 6, 7, 8, 9, . . .

Addition Table

+	0	1	2	3	4	5	6	7	8	9
0	0	1	2	3	4	5	6	7	8	9
1	1			4						
2	2			5						
3	3			6						
4	4			7						
5	5			8						
6	6	7	8	9	10	11	12	13	14	15
7	7									
8	8									
9	9									

You can use the addition table to add.

Example: To add 6 + 3, find where the row for 6 and the column for 3 meet. They meet at 9. What is another way to find the answer? (Use the row for 3 and the column for 6.)

B. Use the finished table above to complete the problems.

1. What four ways are shown in the table to get an answer of 3?

     _____ + _____ = 3   _____ + _____ = 3

     _____ + _____ = 3   _____ + _____ = 3

2. **Discuss** Are the answers to 1 + 2 and 2 + 1 the same? Try switching the order of numbers in other addition facts. Describe what happens. (This is the **commutative property** of addition. The order of the numbers being added doesn't matter. The sum will be the same.)

C. Use the finished table to find combinations that add to the given number. Look for patterns.

3. _____ + _____ = 5
_____ + _____ = 5
_____ + _____ = 5

5. _____ + _____ = 7
_____ + _____ = 7
_____ + _____ = 7
_____ + _____ = 7

7. _____ + _____ = 9
_____ + _____ = 9
_____ + _____ = 9
_____ + _____ = 9
_____ + _____ = 9

4. _____ + _____ = 6
_____ + _____ = 6
_____ + _____ = 6
_____ + _____ = 6

6. _____ + _____ = 8
_____ + _____ = 8
_____ + _____ = 8
_____ + _____ = 8
_____ + _____ = 8

8. _____ + _____ = 10
_____ + _____ = 10
_____ + _____ = 10
_____ + _____ = 10
_____ + _____ = 10

D. Using the finished table, continue looking for combinations to find these larger sums. If you see a pattern, make a note of it.

9. _____ + _____ = 11
_____ + _____ = 11
_____ + _____ = 11
_____ + _____ = 11

11. _____ + _____ = 13
_____ + _____ = 13
_____ + _____ = 13

13. _____ + _____ = 15
_____ + _____ = 15

14. _____ + _____ = 16
_____ + _____ = 16

10. _____ + _____ = 12
_____ + _____ = 12
_____ + _____ = 12
_____ + _____ = 12

12. _____ + _____ = 14
_____ + _____ = 14
_____ + _____ = 14

15. _____ + _____ = 17

16. _____ + _____ = 18

E. Use the finished table to fill in the blanks. Look for patterns.

17. 5 + 7 = _____
5 + _____ = 12
_____ + 7 = 12

19. 3 + _____ = 3
_____ + 0 = 3
3 + 0 = _____

21. 1 + 1 = _____
2 + 2 = _____
3 + 3 = _____
4 + 4 = _____
5 + 5 = _____
6 + 6 = _____
7 + 7 = _____
8 + 8 = _____
9 + 9 = _____

18. 2 + 6 = _____
_____ + 6 = 8
2 + _____ = 8

20. 9 + 1 = _____
_____ + 1 = 10
9 + _____ = 10

What patterns do you see? Explain your reasoning.

Addition Facts

You can use a number line to picture an addition problem in your mind.
Suppose you add 2 + 4. Begin at 2 and move 4 units to 6.

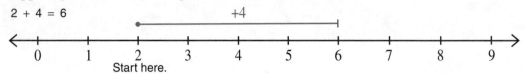

2 + 4 = 6

A. Use the number lines to fill in the blanks.

1.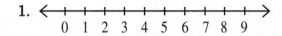

   0 1 2 3 4 5 6 7 8 9

   3 + _____ = 9

4. 0 1 2 3 4 5 6 7 8 9

   _____ + 5 = 7

2. 0 1 2 3 4 5 6 7 8 9

   4 + 4 = _____

5. 0 1 2 3 4 5 6 7 8 9

   2 + 7 = _____

3. 0 1 2 3 4 5 6 7 8 9

   3 + _____ = 7

6. 0 1 2 3 4 5 6 7 8 9

   9 + 0 = _____

As you can see from the addition table and the number line, addition is really a shortcut for counting. You will feel more confident when you add if you memorize the **addition facts.** Here are a few tips:

- Addition can be written in two ways: **horizontally** or **vertically**

$$2 + 3 = 5 \qquad \begin{array}{r} 2 \\ + 3 \\ \hline 5 \end{array}$$

- The order of the numbers does not change the answer.

  Examples: 6 + 3 = 9 and 3 + 6 = 9

- When 0 is added to any number, the number stays the same.

  Examples: 1 + 0 = 1 and 6 + 0 = 6

- When 1 is added to any number, the answer is the next counting number.

  Examples: 6 + 1 = 7 and 10 + 1 = 11

- Some special combinations are numbers that add up to 10.

  Examples: 1 + 9 = 10 and 2 + 8 = 10

- When you add a number to itself, you get an even number.

  Examples: 1 + 1 = 2 and 2 + 2 = 4

B. Practice your addition facts. Practice these problems until you can do them quickly with no errors.

7.
$$8 \\ + 6$$

$$4 \\ + 3$$

$$6 \\ + 3$$

$$1 \\ + 1$$

$$5 \\ + 9$$

$$4 \\ + 9$$

$$3 \\ + 5$$

$$2 \\ + 5$$

$$8 \\ + 7$$

$$3 \\ + 7$$

$$8 \\ + 0$$

8. $6 + 7 =$ _____
$5 + 8 =$ _____
$1 + 5 =$ _____

$4 + 4 =$ _____
$2 + 8 =$ _____
$0 + 2 =$ _____

$7 + 3 =$ _____
$2 + 4 =$ _____
$7 + 1 =$ _____

$8 + 4 =$ _____
$7 + 4 =$ _____
$6 + 5 =$ _____

$5 + 1 =$ _____
$5 + 2 =$ _____
$7 + 5 =$ _____

$5 + 3 =$ _____
$5 + 6 =$ _____
$7 + 9 =$ _____

$0 + 1 =$ _____
$6 + 8 =$ _____
$3 + 2 =$ _____

$8 + 3 =$ _____
$6 + 4 =$ _____
$5 + 7 =$ _____

$1 + 2 =$ _____
$9 + 0 =$ _____
$0 + 0 =$ _____

$9 + 6 =$ _____
$5 + 0 =$ _____
$3 + 9 =$ _____

9.
$$4 \\ + 2$$

$$7 \\ + 6$$

$$4 \\ + 6$$

$$6 \\ + 2$$

$$9 \\ + 4$$

$$7 \\ + 8$$

$$2 \\ + 7$$

$$9 \\ + 9$$

$$9 \\ + 3$$

$$1 \\ + 4$$

$$8 \\ + 2$$

10. $8 + 1 =$ _____
$4 + 1 =$ _____
$2 + 9 =$ _____

$6 + 1 =$ _____
$8 + 5 =$ _____
$6 + 0 =$ _____

$5 + 4 =$ _____
$8 + 9 =$ _____
$9 + 2 =$ _____

$0 + 3 =$ _____
$9 + 5 =$ _____
$3 + 6 =$ _____

$6 + 6 =$ _____
$9 + 7 =$ _____
$4 + 0 =$ _____

$4 + 8 =$ _____
$2 + 7 =$ _____
$1 + 9 =$ _____

$1 + 3 =$ _____
$1 + 7 =$ _____
$3 + 4 =$ _____

$2 + 6 =$ _____
$2 + 1 =$ _____
$7 + 7 =$ _____

$8 + 8 =$ _____
$9 + 1 =$ _____
$3 + 3 =$ _____

$4 + 7 =$ _____
$6 + 9 =$ _____
$1 + 6 =$ _____

Addition Strategies

Because we have a base ten system, special combinations involving 10 are helpful.

Strategy 1. Adding 10 to a one-digit number.
Place a 1 in the tens place next to the number. What does the 1 represent?

$$4 + 10 = 14$$

value of 10 ⌐⌐ value of 4

Try $10 + 3 =$ _____ and $9 + 10 =$ _____.

If you wanted to add 20 to a one-digit number, can you put a 2 in the tens place? Explain.

Strategy 2. Adding 9 to a number.
The answer is 1 less than when you add 10.

$$6 + 9 = 15$$

└ 1 less than 16

Try $9 + 3 =$ _____ and $8 + 9 =$ _____.

Strategy 3. Adding three or more numbers.
Add two numbers first. Then add that sum to the next number.
If possible, look for pairs that add to 10.

$$2 + 8 + 3 = 10 + 3 = 13$$

10

Try $5 + 7 + 3 =$ _____ and $9 + 8 + 1 =$ _____.

A. Practice the addition strategies and facts in these problems.

1. $10 + 6 =$ _____   _____ $+ 10 = 18$ $10 + 1 =$ _____ $10 +$ _____ $= 12$

2. $7 + 9 =$ _____ $9 +$ _____ $= 12$ $5 +$ _____ $= 14$ $8 + 9 =$ _____

3. $3 + 4 + 6 =$ _____ $8 + 7 +$ _____ $= 17$ $5 + 9 + 5 =$ _____

4.
$$\begin{array}{cccc} 10 & 9 & 7 & 8 \\ +\,5 & +\,2 & +\,7 & +\,2 \end{array}$$

B. For more practice, solve the problems.

5.
$$\begin{array}{cccccccc} 8 & 2 & 2 & 6 & 7 & 3 & 5 & 1 \\ 5 & 8 & 4 & 4 & 8 & 7 & 1 & 9 \\ +\,2 & +\,5 & +\,6 & +\,2 & +\,3 & +\,8 & +\,9 & +\,5 \end{array}$$

6. Look at the problems above. Does it matter if the order of the numbers is changed? Explain.

Addition Equations

An **addition equation** is a concise way to write a number sentence. The equation must use a plus sign and an equal sign.

Example: *Eight plus four equals twelve* is written $8 + 4 = 12$

$4 + 8 = 12$ $12 = 4 + 8$ $12 = 8 + 4$ mean the same as $8 + 4 = 12$

C. Write four addition equations using the given numbers.

Use these numbers: 3, 6, and 9.

7. a. _____   b. _____   c. _____   d. _____

Use these numbers: 7, 8, and 15.

8. a. _____   b. _____   c. _____   d. _____

Sometimes one number in the sentence is unknown. A **variable** such as a letter of the alphabet, a box, or a blank can be used in place of the unknown number. If the same variable is used more than once in an equation, it stands for the same number.

Examples: $5 + 2 = $ ▨   In this sentence, the unknown ▨ equals 7.

   $n + 3 = 8$ ***Think:*** what added to 3 equals 8? In this sentence, n represents 5.

   $B + B = 16$ ***Think:*** what added to itself equals 16? In this sentence, the value of B is 8.

D. In each problem, find the value of the variable that makes the equation true.

9. $n + 9 = 17$

   $n = $ _____

11. $4 + G + 9 = 19$

   $G = $ _____

13. $5 + 5 + y = 15$

   $y = $ _____

10. $A + A = 6$

   $A = $ _____

12. $5 + 8 = t$

   $t = $ _____

14. $x + x + 5 = 13$

   $x = $ _____

15. **Write** At the pet shop Todd sold 8 goldfish and 9 guppies. How many fish did he sell that day? Write an addition equation and solve for the variable.

16. **Write** Molly made $5 per hour at her job. With her raise, she will make $7 per hour. How much was Molly's raise? Write an addition equation and solve for the variable.

Adding Larger Numbers

Remember that we can only combine things that are alike. So when you add larger numbers, it's important to line up the digits that have the same *place value*. That way you'll add ones to ones, tens to tens, and so on. If you line up the ones place, the other digits will fall in place.

Let's think in terms of money. To add $412 and $56, think about how many ones, or singles, you have. Add those ones first. Then add the tens together. Now add the hundreds.

Adding Larger Numbers

Example: Line up the digits in the problem $412 + $56. Then add.

Step 1	Step 2	Step 3	Step 4	Step 5
Estimate. Round $412 to $400 and $56 to $60.	Line up the digits in the ones place.	Add the ones.	Add the tens.	Add the hundreds.
$400 + 60 $460	$412 + 56	$412 + 56 8	$412 + 56 68	$412 + 56 $468

The exact answer, **$468,** is close to the estimate of $460. **Estimation** can be helpful in different ways. It can give you an approximate answer or help you check to see if your answer makes sense.

A. **Estimate the answers first, then add. Compare your exact answers with the estimates. The first one is started for you.**

1.
	Estimate				
$34 + 12	$30 + 10	$79 + 20	$802 + 154	281 + 506	$2,345 + 3,104

2. 73 + 6 $483 + $13   507 + 12 $381 + $1,107   636 + 363

B. **For more practice, solve the problems below.**

3.
$46 2 + 351	273 503 + 11	$1,253 + 425	$723 41 + 10	222 242 + 424

4. **Discuss** With others in the class, discuss why you start adding the ones first. Does it make sense to add from right to left?

50

C. Use addition equations (see page 49) to solve the problems.

5. On Monday Paula made 24 copies of the manager's report. On Tuesday she made 120 more copies to be distributed to all employees. How many copies did she make altogether?

6. It is 134 miles from Milwaukee to Chicago. If Emily bought a round-trip ticket between the two cities, what will the total mileage be for her trip?

7. Samly pays $350 per month for rent. His landlord recently told him the rent will go up $35 next year. What will his rent be next year?

8. Jason kept track of the weekend video rentals at Videos for View on the chart below. What was the sum of the rentals for the four-day weekend?

Day	Number of Rentals
Thursday	103
Friday	220
Saturday	310
Sunday	50

Making Connections: Buying a Car

When you buy a car, there is a standard vehicle price. However, the price increases if you include optional equipment. Look at the sticker price list for the new luxury sedan. Then answer the questions.

Luxury Sedan

Standard Price	**$25,300**
Special Console	1,032
V8 Engine	1,033
Auto Trans	NC
Metallic Paint	101
Air Cond	1,032
AM/FM Ster/Cass	+ 300
Vehicle + Options	_____
Destin/Del/Txs	+ 1,000
TOTAL	_____

1. What is the standard vehicle price?

2. Estimate the vehicle + options price.

3. Fill in the exact vehicle + options price. Compare with your estimate.

4. Fill in the total, including destination, delivery charges, and taxes.

Adding with Regrouping

Sometimes when you combine amounts, the sum is large enough for you to create new combinations. For example, if you add $9 and $7, you get $16. Instead of leaving the amount as 16 ones or singles, think of it as 1 ten-dollar bill and 6 ones.

When adding, if the sum of the digits in any column is 10 or more, write the last digit beneath the column and **regroup** (carry) the extra digit to the next place column. In the paragraph above, the regrouped amount would be the 1 ten. You would then add that ten to any other tens being added.

Adding and Regrouping

Example: $389 + $57

Step 1	Step 2	Step 3	Step 4
Estimate first.	Add the ones. Write 6 in the ones column. Regroup the 1 ten to the tens place.	Add the tens. Write 4 in the tens column. Regroup the 1 hundred to the hundreds place.	Add the hundreds (including the regrouped amount).

$400	¹	¹ ¹	¹ ¹
+ 60	$389	$389	$389
$460	+ 57	+ 57	+ 57
	6	46	$446

The answer, **$446,** is close to the estimate of $460.

A. Add, regrouping if necessary. Estimate first.

	Estimate				
1. $37	$40	813	$648	$2,349	686
+ 46	+ 50	+ 58	+ 257	+ 943	+ 486

2. 59 + 17 $841 + $73   8,156 + 175 $483 + $19 + $27

B. For more practice, add the following numbers.

3. $624 + $38   5,680 + 17   985 + 47 $7 + $18 + $346

C. Solve the problems below. You may use equations (see page 49) to set up the problems before solving.

4. At Warehouse, Inc., the employees collected toys to give to children during the holiday season. They collected 46 dolls, 18 toy trucks, 39 puzzles, and 53 games. How many total presents did they collect?

5. Doug earns $1,750 per month and Sandra earns $2,364 per month. How much do they earn together?

6. The 5 starting offensive linemen on the Hometown High School football team weigh 186 pounds, 225 pounds, 256 pounds, 192 pounds, and 168 pounds.

   a. What is their combined weight?

   b. If you increase the line with two more players at 195 pounds and 187 pounds, what is the new total?

7. Attendance figures at the local art exhibit were 428 people on Friday, 376 on Saturday, and 721 on Sunday. In all, how many attended?

Making Connections: Maps and Distances

The mileage between cities is shown on the map.

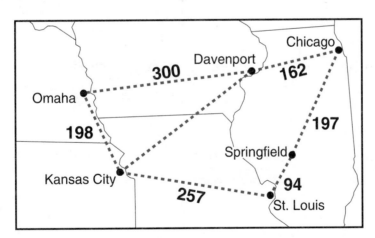

Use the map to answer the following questions.

1. What is the distance from Chicago to Kansas City by way of St. Louis? Is this the shortest way to go?

2. If you live in Chicago and visit relatives in Omaha, how far is the trip there and back?

3. A driving trip around the Midwest includes all the cities shown on the map. What is the total mileage of the trip?

Finding Perimeter

Geometry is the study of shapes. As you look around, you see shapes such as squares, rectangles, and triangles. These flat figures are called **polygons.** *Poly* means many and *gon* means sides. Polygons also include pentagons (5 sides) and decagons (10 sides).

The sides of polygons are lines. The length of a line can be measured with a ruler, tape measure, or yardstick. The units on these measuring tools are either **English** or **metric.**

English Measures of Length	Metric Measures of Length
1 inch (in.) = 1 inch	1 millimeter (mm) = 1 mm
1 foot (ft.) = 12 inches	1 centimeter (cm) = 10 mm
1 yard (yd.) = 3 feet	1 meter (m) = 100 cm
1 mile (mi.) = 5,280 feet	1 kilometer (km) = 1,000 m

A. Find the total length of each line.

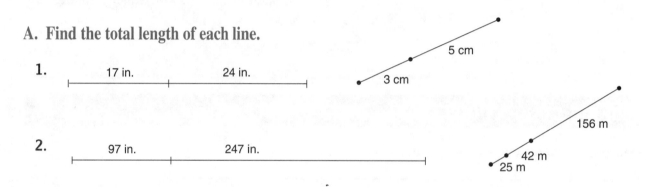

1. 17 in.   24 in.   3 cm   5 cm

2. 97 in.   247 in.   25 m 42 m 156 m

Some common geometric shapes are shown below. (*Note:* A **right angle** measures 90°.)

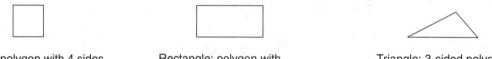

Square: polygon with 4 sides of equal length and 4 right angles

Rectangle: polygon with opposite sides of equal length and 4 right angles

Triangle: 3-sided polygon

B. Name the figures shown below.

3.

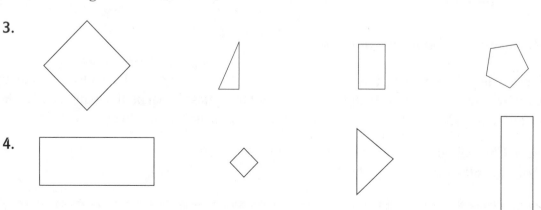

4.

Working with Perimeter

Perimeter is the distance around a polygon. The variable *P* usually stands for perimeter.

Finding Perimeter

To find perimeter, add the lengths of the sides of the polygon.

Example 1: Find the perimeter of the 4-sided polygon. Is it a rectangle?

P = 8 in. + 5 in. + 9 in. + 18 in.

P = 40 in.

It is *not* a rectangle because it does not have equal opposite sides or four right angles.

Example 2: Find the perimeter of the square. Why is the length of only one side given?

P = 6 cm + 6 cm + 6 cm + 6 cm

P = 24 cm

☐ 6 cm

You only need one side's length since *a square's sides are equal.*

For another look at formulas, turn to page 205.

C. Find the perimeter of each figure shown below.

5.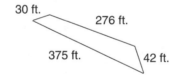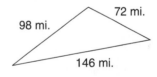

6.   rectangle   square      rectangle

D. Use your addition and geometry skills to solve these problems.

7. What is the distance around the rectangular city block shown below?

   220 yd.
   ☐ 110 yd.

8. The Kirks fenced in their 6-acre pasture. Find the perimeter of their rectangular pasture.

   ☐ 726 ft.
   2,160 ft.

9. How many feet of binding are needed to bind the edges of the sail shown below?

10. **Draw** A baseball diamond is square. What is the distance around the bases if it is 90 feet from first to second base? Draw a diagram to show your answer.

Mixed Review

A. Use the addition facts to answer the problems.

1. Fill in the blanks to complete the sums.

   a. 2 + 3 = _____   4 + _____ = 10   _____ + 6 = 11

   b. 7 + 4 = _____   _____ + 5 = 13   7 + _____ = 15

   c. 8 + 0 = _____   9 + _____ = 18   6 + 3 = _____

   d. 9 + 6 = _____   _____ + 1 = 10   8 + 3 = _____

2. Write eight combinations of numbers that add to 16.

    _____ + _____ = 16   _____ + _____ = 16

    _____ + _____ = 16   _____ + _____ = 16

    _____ + _____ = 16   _____ + _____ = 16

    _____ + _____ = 16   _____ + _____ = 16

3. Add the numbers below.

   a.
$$\begin{array}{r} 5 \\ 7 \\ +\,5 \\ \hline \end{array} \qquad \begin{array}{r} 6 \\ 3 \\ +\,4 \\ \hline \end{array} \qquad \begin{array}{r} 7 \\ 2 \\ +\,9 \\ \hline \end{array}$$

   b.
$$\begin{array}{r} 8 \\ 6 \\ +\,2 \\ \hline \end{array} \qquad \begin{array}{r} 7 \\ 7 \\ +\,5 \\ \hline \end{array} \qquad \begin{array}{r} 6 \\ 7 \\ +\,3 \\ \hline \end{array}$$

4. Perform the addition operation on these numbers.

$$\begin{array}{r} 48 \\ +\,26 \\ \hline \end{array} \qquad \begin{array}{r} 384 \\ +\,25 \\ \hline \end{array} \qquad \begin{array}{r} 487 \\ +\,13 \\ \hline \end{array} \qquad \begin{array}{r} 526 \\ +\,39 \\ \hline \end{array} \qquad \begin{array}{r} 2{,}341 \\ +\,293 \\ \hline \end{array}$$

5. Add these numbers.

    56 + 3 + 28   476 + 25 + 64   98 + 23 + 15

B. Solve the problems below.

6. Find the value of the variable in each problem.

    8 + 6 = N   C + 4 = 12   5 + 10 = V   9 + Y = 13

    N =   C =   V =   Y =

7. Write an addition equation using the numbers 14, 8, and 6.

8. On her vacation Diane took a lot of pictures with her camera. She used one 24-exposure roll of film and two 36-exposure rolls. How many total exposures did she use?

C. Use the information about Rosa to answer these problems.
Rosa has designed 3 pillows to sell at the craft show.

 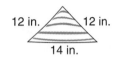 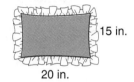

9. If she trims the square pillow with white lace, how many inches long will the lace trim be?

10. If Rosa trims two triangular pillows with navy blue binding, find the total length of the navy trim.

11. Find the total length of the ruffle on the rectangular pillow.

D. Review some addition strategies. Discuss these questions with a partner.

12. When you add a number to itself, is your answer an odd or even number? Support your choice with a few examples.

13. a. What happens when you add 0 to a number?

    b. What happens when you add 10 to a one-digit number?

    c. What happens when you add 9 to a number?

14. Name the five pairs of numbers that add to 10.

15. Discuss Are there any pairs of numbers whose sums are hard to remember? Write them and discuss strategies you might use to remember them.

Adding Thousands

When adding large numbers, you may need to regroup several times.

Adding Thousands

Example 1: 6,285 + 95,784

Estimate first.

Add the numbers from right to left. Be sure to write the regrouped number at the top.

```
    10,000
 + 100,000
   110,000
```

```
    ¹ ¹
   6,285
    ¹
 + 95,784
  102,069
```

The answer, **102,069,** is close to the estimate of 110,000.

Example 2: 287,643 + 95,188 + 14,999

Estimate first.

```
   300,000
   100,000
 +  15,000
   415,000
```

```
   ¹¹¹ ²²
   287,643
    95,188
 +  14,999
   397,830
```

Notice that the ones total 20. Write 0 in the ones place and regroup the 2 tens.

The answer, **397,830,** is close to the estimate of 415,000.

A. Practice adding these large numbers. Estimate first.

	Estimate		
1. 56,783	*60,000*	243,176	6,845,199
+ 8,104	*+ 8,000*	+ 595,123	+ 167,381

2. 38,176   1,284,565   571,288
    49,287   100,218   143,044
+ 54,399   + 391,309   + 502,311

B. For more practice, solve these problems.

3. What is the difference in elevation between the highest point in North America, Mount McKinley in Alaska, and the lowest point, Death Valley, California?

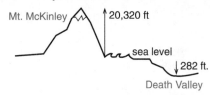

Mt. McKinley   ↑20,320 ft

sea level

↓282 ft.

Death Valley

4. Henri's wages are based on his total sales. If he sells the cars whose prices are listed below, what will his sales total be?

$9,899 $14,565 $28,394 $17,656

C. **Practice your addition skills with this cross number puzzle.**
   (Hint: First fill in the answers you can do quickly; see 32 down.)

Across

1. 29 + 9

3. 868 + 728

7. 285 + 150

8. 43,288 + 24,107

10. 2,813 + 2,478

12. 200 + 40 + 1

13. 12,890 + 17,724

15. 26 + 6

16. 58 + 4

17. 86 + 72 + 21

19. 7 + 8

21. 6 + 7

22. 500 + 40

23. 27 + 9

24. 9 + 9

25. 31,109 + 41,239

28. 100 + 29

30. 600 + 509 + 400

31. 37,570 + 39,941

34. 190 + 105

35. 2,094 + 910

36. 8 + 4

Down

1. 330,003 + 2,020

2. 1,478 + 7,118

3. 6 + 10

4. 47 + 10

5. 9,000 + 300 + 22

6. 396 + 298

7. 26,811 + 18,550

9. 47 + 4

11. 5 + 6

14. 400 + 14

15. 15 + 15 + 9

18. 245 + 462

19. 5,092 + 30,000 + 99,000

20. 56,230 + 665

22. 20 + 8 + 30

24. 1,000 + 950

26. 18 + 3

27. 1,759 + 1,762

28. 8 + 9

29. 150 + 123

32. 6 + 4

33. 7 + 7

Understanding Word Problems

Word problems describe real-life situations. You can use math skills to solve these problems. A first step to understanding a word problem is to identify the question that is being asked. Underline the question.

Identifying the Question

Example 1: During the first three weeks of the month, Shay Sporting Goods had sales totals of $4,256 and $3,472 and $2,895. <u>What were the combined sales?</u>

Example 2: <u>Find the new enrollment figures</u> for the park district summer programs if there is an increase of 1,256 people from the 15,728 people enrolled last year.

A. Underline the question in each word problem. Do not solve the problems yet.

1. Ben buys office supplies for his company. Envelopes come in boxes of 500, 750, and 1,000. How many envelopes should he order if the secretary needs 375 envelopes for a brochure and 289 envelopes for correspondence?

2. Find the total number of lunches served on Monday, Wednesday, and Friday at the senior center for the week shown below.

Day	Lunches Served
Monday	43
Tuesday	41
Wednesday	39
Thursday	56
Friday	62
Saturday	40

3. Combine 4 cups of cranberry juice, 2 cups of apple juice, and 8 cups of ginger ale to make a delicious punch. How many cups of punch will a double recipe make?

4. At the community fund-raiser, pledges were received for $56, $185, $350, and $145.
   a. Altogether, how much money was pledged?
   b. Did the community reach its goal of $1,000 in pledges?

5. **Write** Write your own addition problem and underline the question.

Word Problem Strategies

Here are several useful problem-solving strategies.

Strategy 1. *Personalize the problem.* Think about how you would solve the problem in your life.

Example 1: Find the total hours Rattana worked last week. She worked 6 hours a day on 3 days, 8 hours on Thursday, and 7 hours on Tuesday. *Can you relate this to a situation in your life that takes a lot of time?*

M	T	W	T	F

Strategy 2. *Draw a picture.* When geometric shapes are a part of the problem, it often helps to sketch the figure to organize your thoughts.

Example 2: How many feet of fencing will you need to enclose a rectangular garden 36 feet long and 18 feet wide? *What drawing would you make?*

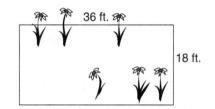

Strategy 3. *Write a number sentence.* A number sentence can help you translate the problem from words to a math operation.

Example 3: Alex drove 186 miles the first day, 283 miles the second day, and only 98 miles the last day. How far did he drive? *What number sentence do you see in this problem?*

$$186 + 283 + 98 = x$$

Strategy 4. *Make a list.* A list can help you organize the information given in a problem. Look at Example 4 below.

Strategy 5. *Guess and check.* Sometimes you have to try more than one way to solve a problem. Make sure your answer makes sense.

Example 4: Wade has monthly expenses that include $76 for electricity, $38 for phone, $575 for rent, $108 for insurance, $40 for laundry, $150 for transportation, and $225 for food and entertainment. Which combination of expenses can he pay with an $825 paycheck?

First make a list of the expenses. There are many possible combinations. Use estimation to help you make a sensible guess. Check your work with the actual addition.

B. Solve the examples above, and discuss the strategies you used.

C. Solve the problems in Part A on page 60.

Adding on a Calculator

The **calculator** has become commonplace in our daily lives. Although not all calculators are alike, they all have keys similar to the ones shown below. We will be using the following keys in this book:

⬚ On/C   Press the On/Clear key to turn the power on and clear the display. A zero and decimal point appear in the display. *You must press the clear key to begin any new calculation.*

⬚ 1 2 3 4 5 6 7 8.   The display has enough room to show about eight digits and a decimal point. Most calculators don't use or show commas.

CE   Press the Clear Entry key to clear only the last number entered, not the whole problem.

1 2 3
4 5 6
7 8 9
0 ·
   Press the digit keys and the decimal point key to enter numbers.

÷ ×
− +
   Press an operation key to perform the necessary operation.

=   Press the equal sign to complete the calculation.

> ► **Using a Calculator**
>
> The basic pattern to enter a problem is
> - press On/Clear
> - enter the first number
> - press the operation symbol
> - enter the next number
> - press the equal sign

Example: Always estimate an answer before you begin.
(*Note:* ≈ means "is approximately equal to.")
$48 + 37 \approx 50 + 40 = 90$
To add $48 + 37$, press the keys in the following order.

Key in:	Your display reads:
On/C	0.
4 8	48.
+	48. ← The plus sign does not display.
3 7	37.
=	85. ← total

Your answer, **85,** is close to the estimate of 90.

Entering Several Numbers

Example: 257 + 35 + 789

Estimate first: 257 + 35 + 789 ≈ 250 + 50 + 800 = 1,100

Key in:	Your display reads:
On/C	0.
2 5 7	257.
+	257.
3 5	35.
+	292. ← subtotal
7 8 9	789.
=	1081. ← total

A. Get to know your calculator. Compare and discuss your answers with a group.

1. Press the On/C key. What is displayed?

2. Press the digit keys. How many digits can be displayed at once?

3. Press the key to clear the display. What shows on the display?

4. **a.** Enter 23 + 56. What is displayed?
   b. Press CE , then 5 7 . What is displayed?
   c. Press the = . What is displayed? What happened?

The calculator is a valuable tool. However, if you enter a wrong number or symbol, you will get a wrong answer. Estimation can help determine if you entered the correct symbol or number of zeros.

B. Find an estimate. Then use a calculator to add the numbers.

5. 89 + 243   Estimate: _____ + _____ = _____   Answer: _____

6. 2,356 + 7,084   Estimate: _____ + _____ = _____   Answer: _____

7. 53,941 + 47,821   Estimate: _____ + _____ = _____   Answer: _____

C. Use a calculator to add the numbers. Be sure to estimate first.

8.	438	9.	46	10.	2,349,851
	296		23		+ 3,777,265
	157		98		
	740		99		
	+ 937		+ 74		

Adding Dollars and Cents

When you add money, keep these tips in mind:

Tip 1. When adding dollars, line up like place values.

    Example: $48 + $7
$$\begin{array}{r} \$48 \\ +\ 7 \\ \hline \$55 \end{array}$$

Tip 2. When adding cents, you may use the cent symbol ¢ if the sum is less than $1. Estimate to see if the sum is less than $1.

    Example: 48¢ + 25¢
$$\begin{array}{r} 48¢ \\ +\ 25¢ \\ \hline 73¢ \end{array}$$

Tip 3. If the sum of the cents is more than $1, use the dollar symbol $ and a decimal point. Line up the points.

    Example: 75¢ + 28¢
$$\begin{array}{r} \$0.75 \\ +\ 0.28 \\ \hline \$1.03 \end{array}$$

Tip 4. When adding dollars and cents, line up the decimal points.

    Example: $12 + $1.82 (remember $12 is $12.00)
$$\begin{array}{r} \$12.00 \\ +\ 1.82 \\ \hline \$13.82 \end{array}$$

A. Add the amounts below.

1. $36 + $24 $250 + $85   75¢ + 16¢

2. $2,346 + $925   86¢ + 56¢   72¢ + $1.48

3. $8.57 + $9.10 $483.50 + $16.95 $18 + $1.96

B. Add the following money amounts. You may use a calculator.

4. $75 + $9.24 $375.50 + $43.65   84¢ + 36¢

5. 18¢ + 42¢ $99 + $38.45 $27 + $6

C. Solve the following problems. Estimate your answers first.

6. In 1994 José earned $26,500. If he received a year-end bonus of $1,800, what were his total 1994 earnings?

7. Mike and Kim bought a bookshelf stereo system for $479 plus 8% tax. What was their total cost if the tax was $38.32?

8. Hal bought a cordless screwdriver for $25.99 and a new hammer for $15.95. He paid $3.36 in sales tax. What was his total bill?

9. Sales at the Denver Deli were $675.28 on Saturday and $429.55 on Sunday. Find the total weekend sales.

Making Connections: Ordering from a Catalog

Find the total bill on each catalog order. Find a *merchandise total* first. Then add *shipping and handling charges* to find the final bill.

1.

Qty	Item Number	Description	Price Each	Total
2	34207	mini-chopper	$19.95	$39.90
1	14752	wok	25.99	25.99
2	30521	wooden utensils	3.49	6.98
Kitchen & Cooks			Merchandise Total	
			Shipping & Handling	3.95
			Total	

2. **Camping Warehouse**

Qty	Item Number	Description	Price Each	Total
1	42111	tent	$298.99	$298.99
1	21475	sleeping bag	58.25	58.25
1	74281	lantern	19.99	19.99
			Merchandise Total	
			Shipping & Handling	7.50
			Total	

Estimating Costs

Addition is a valuable skill to use when shopping. You can use it to estimate the cost and then use it again to find the exact cost.

When you shop, the total bill is the sum of the prices of the items plus any sales tax. To practice finding the total bill, we will use an 8% tax table. To examine estimating, let's see what an 8% tax really means.

8% Tax Means
8¢ tax on each $1
80¢ tax on each $10
$8 tax on each $100
$80 tax on each $1,000
$800 tax on each $10,000

For instance, an item priced at $100 would cost a total of $100 + $8 or $108. An item priced at $200 would cost $200 + $16 or $216.

A. Use the table above to answer and discuss these questions.

1. What is the total bill if you buy an item for
   a. $10?
   b. $20?

2. What is the total bill if you buy something for
   a. $1,000?
   b. $4,000?

To estimate the total bill	**To find exact cost**
round up each item's price	add all items
add to get a subtotal	select exact tax
estimate the tax	add to get total bill
add to get total estimate	

Suppose you want to buy the items below. First estimate your total bill. When shopping, *round up* your estimates to be sure you have enough money. If the estimate sounds good, calculate the exact cost.

Turtleneck: $12.98

Rugby shirt: $29.95

Jeans: $26.50

Estimate: $13 + $30 + $30 = **$73**   Exact: $12.98 + $29.95 + $26.50 = **$69.43**

**B. Solve these problems with a partner. Discuss your estimates.
Are they reasonable? Discuss the tax.**

First estimate each total bill. Then find the exact amount of the total bill. You may use a calculator to find the exact amounts.

3. At the grocery store
(*Note:* Some states charge sales tax on food.)

		Exact	Estimate
1 whole chicken		$3.98	_____
5 pounds potatoes		1.49	_____
1 head lettuce		.98	_____
1 tray tomatoes		1.29	_____
1 pound frozen peas		.89	_____
6-pack cola		1.69	_____
1 gallon milk		2.59	_____
1 pound wheat bread		1.29	_____
Subtotal		_____	_____
Tax		_1.14_	_____
Total		_____	_____

4. At the hardware store

		Exact	Estimate
smoke alarm		$3.99	_____
cordless drill		39.00	_____
1 pound screws		4.78	_____
table saw		165.99	_____
Subtotal		_____	_____
Tax		_17.10_	_____
Total		_____	_____

5. At the discount store

		Exact	Estimate
watch		$44.99	_____
wallet		9.88	_____
gold chain		149.99	_____
bread maker		179.99	_____
cordless phone		78.88	_____
Subtotal		_____	_____
Tax		_37.10_	_____
Total		_____	_____

Unit 2 Review

A. Answer the problems below.

1. Write the missing number in each number pattern:

    a. 5, 10, _____, 20, 25, . . .   **b.** 4, 8, 12, 16, _____, . . .   **c.** _____, 6, 9, 12, . . .

2. Match the numbers in the first column with the words in the second column.

    _____ 7,028   **a.** seven hundred twenty-eight

    _____ 728   **b.** seventy thousand two hundred eighty

    _____ 70,280   **c.** seven thousand twenty-eight

    _____ 7,208   **d.** seven thousand two hundred eight

3. Compare the numbers using =, >, or <.

    182 ___ 376   4,506 ___ 4,056   8 + 7 ___ 7 + 8

4. Match the dates.

    _____ 4/9/94   **a.** September 9, 1994

    _____ 9/9/94   **b.** April 4, 1994

    _____ 9/4/94   **c.** April 9, 1994

    _____ 4/4/94   **d.** September 4, 1994

5. Use digits to write the time next to the clocks below.

6. Identify the points on the number line.

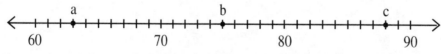

    a. _____   **b.** _____   **c.** _____

7. Round each number to the value indicated.

    a. 327 to the nearest hundred _____

    b. $67.98 to the nearest $10 _____

B. Solve the addition problems.

8.
$$
\begin{array}{r} 9 \\ +\ 7 \\ \hline \end{array}
\qquad
\begin{array}{r} 16 \\ +\ 23 \\ \hline \end{array}
\qquad
\begin{array}{r} 24 \\ +\ 96 \\ \hline \end{array}
\qquad
\begin{array}{r} 283 \\ +\ 175 \\ \hline \end{array}
\qquad
\begin{array}{r} 3,249 \\ +\ 6,872 \\ \hline \end{array}
$$

9. 8 + 8   14 + 9 + 28   576 + 39 + 644

10.

275,365	67,250	$65.24	$36.56
+ 891,003	+ 30,412	+ 18.97	+ 8.80

11. a. $N + 9 = 16$ **b.** $t + t = 18$ **c.** $23 + 18 = y$

   $N = $ _____ $t = $ _____ $y = $ _____

C. Solve the word problems below. Be sure to identify the question first. (Underline it.) When you complete the problem, ask yourself, "Does the answer make sense?"

12. Tracey wants to put wood molding around the ceiling in her square family room. If the length of one wall is 18 feet, how many feet of molding does she need?

13. Magdalena has $65 to spend on holiday gifts. Can she afford to buy a scarf for $12.95, a tape player for $19.95, and a blouse for $25? The tax is $4.63.

14. Find the subtotal and total on the sales receipt.

Claudia's Cosmetics

perfume	$24.99
hair dryer	19.99
manicure set	8.95
Subtotal	_____
Tax	_4.27_
Total	_____

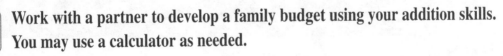

Working Together

Work with a partner to develop a family budget using your addition skills. You may use a calculator as needed.

The Lytle family is trying to decide how to budget their $1,800 monthly income. The chart below shows three possible choices. For each choice, decide how you would budget the remaining money. Be sure each budget totals $1,800.

In the last column, create a new budget for the $1,800.

	Choice 1	Choice 2	Choice 3	Your Choice
Rent	$525	$525	$675	$_____
Utilities	145	190	255	_____
Insurance	180	185	190	_____
Food	275	325	350	_____
Transportation	250	250	250	_____
Miscellaneous	_____	_____	_____	_____
Entertainment	_____	_____	_____	_____
Savings	_____	_____	_____	_____

Subtraction

Skills

Subtraction facts

Subtracting numbers

Regrouping

Subtracting money

Tools

Number line

Calculator

Problem Solvers

Subtraction strategies

Subtraction equations

Deciding to add or subtract

Applications

Checking a paycheck

Making change

Coupons! Coupons! You often see cents-off coupons in newspapers, magazines, and sales flyers. What do you do with them? You can use coupons to reduce the price of the item you're buying. What operation do you use to find the new price? That's right: **subtraction.**

Subtraction is the opposite operation of addition. Addition combines things, and subtraction separates things.

Subtraction is the operation you use to do one of two things:

1. take away an amount to get a smaller amount

2. find the difference between two numbers to make a comparison

As you work with subtraction, rely on the addition skills you learned in Unit 2. Used together, addition and subtraction will help you solve many problems.

When Do I Subtract?

The **minus sign** − is the symbol for subtraction.

Some words or phrases we use with subtraction are listed below.

- take away
- find the amount of change
- how much is left
- what remains

- minus
- find the difference
- how much more or less
- how much longer or shorter

Give an example of when you use subtraction in these settings:

At home: _____

_____

_____

At work: _____

_____

_____

At school: _____

_____

_____

Shopping: _____

_____

_____

Have you had trouble with subtraction in the past? Think of some problems that seem difficult to you. Discuss these with a partner.

Talk About It

When you shop, you often compare prices. When you compare, you subtract. For example, buying a used car requires you to compare prices and conditions of the cars before you make your choice.

Explain how you might use subtraction to compare two cars. For example, you might ask yourself, "Which one has more miles on it? How many more?" What other questions come to mind? Discuss this with a partner.

Subtraction Strategies

As you set up a subtraction problem, keep three things in mind:

- The original number, or starting amount, is written first.
- The minus sign − is written next.
- The number being subtracted is written last.

As with addition, the problem can be set up either

horizontally   or   **vertically**

$$9 - 5 = 4$$

$$\begin{array}{r} 9 \\ -\ 5 \\ \hline 4 \end{array}$$

Subtraction is the opposite operation of addition. You can use the addition table (on page 44) in reverse to help you learn **subtraction facts.**

+	0	1	2	3
0	0	1	2	3
1	1	2	3	4
2	2	3	4	5

Given the fact that $2 + 3 = 5$, the related subtraction facts are $5 - 2 = 3$ and $5 - 3 = 2$.

A. **Write two related subtraction facts for each addition fact. The first one is started for you.**

1. $3 + 1 = 4$ 　　　　 $2 + 4 = 6$ 　　　　 $7 + 3 = 10$
　　$4 - 1 =$
　　$4 - 3 =$

2. $3 + 4 = 7$ 　　　　 $2 + 6 = 8$ 　　　　 $6 + 3 = 9$

3. $4 + 7 = 11$ 　　　　 $6 + 8 = 14$ 　　　　 $7 + 5 = 12$

A **number line** can help you visualize the subtraction operation.

To picture $8 - 3$

- Draw a line from 0 to 8.
- Then move in reverse 3 units.
- Find the answer of 5.

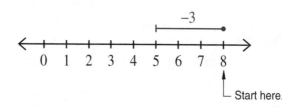

B. Write a subtraction problem to match each number line.

4. a.

b.

c.

You should memorize the subtraction facts as you did with the addition facts. The addition table on page 44 and the number line can help you learn the facts. The tips below will also help you.

Subtraction Tips

Tip 1. Subtracting 0 from a number does not change the number.

$8 - 0 = 8$ $423 - 0 = 423$

Tip 2. Subtracting 1 from a number leaves an answer of 1 less than the number.

$8 - 1 = 7$ $423 - 1 = 422$

Tip 3. Subtracting a number from itself results in an answer of 0.

$8 - 8 = 0$ $423 - 423 = 0$

Tip 4. You can check your answer in subtraction by adding your answer to the number you subtracted.

$12 - 3 = 9$ because $9 + 3 = 12$

C. Test your knowledge of subtraction on the problems below.

5. a.
$$\begin{array}{r} 9 \\ -\ 0 \\ \hline \end{array} \qquad \begin{array}{r} 6 \\ -\ 1 \\ \hline \end{array} \qquad \begin{array}{r} 5 \\ -\ 5 \\ \hline \end{array}$$

b.
$$\begin{array}{r} 3 \\ -\ 0 \\ \hline \end{array} \qquad \begin{array}{r} 4 \\ -\ 1 \\ \hline \end{array} \qquad \begin{array}{r} 7 \\ -\ 7 \\ \hline \end{array}$$

6. $8 - 8 = $ _____ $7 - $ _____ $ = 6$ $5 - $ _____ $ = 5$

$8 - $ _____ $ = 0$ $7 - 1 = $ _____   _____ $ - 0 = 5$

_____ $ - 8 = 0$ _____ $ - 1 = 6$ $5 - 5 = $ _____

7. Write four related addition and subtraction facts using 5, 8, and 13.

For another look at the addition table, turn to page 204.

Subtraction Facts

Now it's time to work with the subtraction strategies and tips you've learned. When solving these problems, keep in mind that subtraction and addition are opposite operations.

A. Practice the basic subtraction facts in these problems until you can do them quickly with no errors.

1. $8 - 3 = $ _____ $0 - 0 = $ _____ $9 - 4 = $ _____ $9 - 9 = $ _____

2. $5 - 0 = $ _____ $9 - 8 = $ _____ $3 - 3 = $ _____ $7 - 4 = $ _____

3. $6 - 1 = $ _____ $6 - 5 = $ _____ $6 - 2 = $ _____ $8 - 2 = $ _____

4. $7 - 0 = $ _____ $3 - 2 = $ _____ $5 - 1 = $ _____ $6 - 0 = $ _____

5. $6 - 6 = $ _____ $4 - 4 = $ _____ $5 - 4 = $ _____ $6 - 4 = $ _____

6. $5 - 2 = $ _____ $9 - 2 = $ _____ $2 - 2 = $ _____ $7 - 7 = $ _____

7. $5 - 5 = $ _____ $1 - 1 = $ _____ $4 - 1 = $ _____ $7 - 1 = $ _____

8. $7 - 6 = $ _____ $2 - 1 = $ _____ $9 - 0 = $ _____ $7 - 5 = $ _____

9. $1 - 0 = $ _____ $8 - 4 = $ _____ $9 - 1 = $ _____ $4 - 2 = $ _____

10. $4 - 3 = $ _____ $3 - 1 = $ _____ $7 - 3 = $ _____ $2 - 0 = $ _____

11. $8 - 5 = $ _____ $5 - 3 = $ _____ $4 - 0 = $ _____ $8 - 8 = $ _____

12. $6 - 3 = $ _____ $8 - 7 = $ _____ $8 - 3 = $ _____ $7 - 2 = $ _____

13. $8 - 1 = $ _____ $3 - 0 = $ _____ $9 - 7 = $ _____ $9 - 5 = $ _____

14. $8 - 6 = $ _____ $9 - 3 = $ _____ $9 - 6 = $ _____ $8 - 0 = $ _____

B. Practice the subtraction facts below until you can do them quickly with no errors.

15. $18 - 9 = $ _____ $11 - 8 = $ _____ $12 - 8 = $ _____ $16 - 7 = $ _____

16. $17 - 8 = $ _____ $10 - 1 = $ _____ $13 - 9 = $ _____ $11 - 6 = $ _____

17. $17 - 9 = $ _____ $13 - 4 = $ _____ $12 - 3 = $ _____ $13 - 5 = $ _____

18. $13 - 7 = $ _____ $12 - 5 = $ _____ $14 - 7 = $ _____ $15 - 8 = $ _____

19. $11 - 9 = $ _____ $14 - 8 = $ _____ $12 - 4 = $ _____ $15 - 7 = $ _____

20. $11 - 2 = $ _____ $11 - 5 = $ _____ $10 - 7 = $ _____ $13 - 6 = $ _____

Subtraction Equations

A **subtraction equation** is a mathematical sentence that takes one number away from another number or finds the difference between two numbers. The sentence uses a *minus sign* and an *equal sign*.

Writing Subtraction Equations

Example 1: Seven take away three equals four is $7 - 3 = 4$.

Example 2: The difference between 5 and 8 is $8 - 5 = 3$.
(Remember, the largest number must be written first.)

C. Write the following sentences as subtraction equations.

21. Twelve take away seven is five.

22. Subtract five from nine to get four.

23. The difference between eleven and four is seven.

24. Eight is two greater than six.

25. Three is seven less than ten.

D. Find the value of the variable (letter) in each equation below.

26. $10 - 3 = r$
$r =$

$Q - 4 = 9$
$Q =$
Hint: _____ $- 4 = 9$

$b - b = 0$
$b =$

27. $N - 9 = 6$
$N =$

$12 - m = 12$
$m =$

$11 - c = 2$
$c =$

E. Decide whether the following equations are true or false. For each question, put a check (✓) under the correct column.

True   False

_____ _____ 28. $3 + 5 = 16 - 8$

_____ _____ 29. $9 - 3 = 2 + 5$

_____ _____ 30. $7 - 7 = 4 - 4$

_____ _____ 31. $12 - 11 = 0 + 1$

_____ _____ 32. $3 + 7 = 12 - 3$

True   False

_____ _____ 33. $16 - 5 = 4 + 7$

_____ _____ 34. $8 - 2 = 15 - 9$

_____ _____ 35. $9 + 6 = 18 - 5$

_____ _____ 36. $13 - 7 = 12 - 6$

_____ _____ 37. $8 + 3 = 18 - 8$

Subtracting Larger Numbers

When you subtract larger numbers, it is important to line up the digits that have the same place value. Like addition, the subtraction operation can be performed only on things that are alike.

If you line up the ones place, the other digits will fall into place. Subtract from right to left, starting with the ones place.

Subtracting Larger Numbers

Example: 386 *yes* votes – 245 *no* votes

Step 1	Step 2	Step 3	Step 4
Line up the digits in the ones place.	Subtract the ones.	Subtract the tens.	Subtract the hundreds.
386 – 245	386 – 245 1	386 – 245 41	386 – 245 **141** more *yes* votes

---

Estimate when you subtract to make sure your answer is sensible. You could round each number to the same place value.

Example: Round 386 to 390 and 245 to 250.
Then subtract the rounded numbers.
390 – 250 = 140

The estimate, 140, is very close to the actual answer of **141** *yes* votes.

A. Estimate first. Then subtract the numbers below. Compare your exact answers with your estimates.

	Estimate				
1.	38 *40* – 12 *– 10*	47 – 25	589 – 453	741 – 530	9,846 – 2,736

2. 87 – 7   63 – 31   529 – 18   672 – 421

3. 744 – 221   689 – 88   365 – 23   565 – 141

Zeros in Subtraction

When you subtract a number from itself, the result is zero. If zero is the lead digit in the answer, do not write it in the answer.

Example:
$$
\begin{array}{r}
283 \\
-\ 271 \\
\hline
12
\end{array}
$$

└ Don't write the zero.

If zero is an inside digit or the end digit in the answer, it is an important **placeholder.** You must write the placeholder (zero) in the answer.

Examples:
$$
\begin{array}{r}
583 \\
-\ 281 \\
\hline
302
\end{array}
\qquad\qquad
\begin{array}{r}
486 \\
-\ 356 \\
\hline
130
\end{array}
$$

└ placeholder; no tens   └ placeholder; no ones

B. Use the second example above to discuss these questions.

4. Describe a placeholder. Why are placeholders necessary?

5. Do the zeros in the answers above mean the same thing? Why or why not?

C. Keep the examples in mind as you practice the problems.

6.
$$
\begin{array}{r} 576 \\ -\ 246 \\ \hline \end{array}
\qquad
\begin{array}{r} 398 \\ -\ 377 \\ \hline \end{array}
\qquad
\begin{array}{r} 856 \\ -\ 854 \\ \hline \end{array}
\qquad
\begin{array}{r} 7{,}349 \\ -\ 241 \\ \hline \end{array}
\qquad
\begin{array}{r} 5{,}921 \\ -\ 2{,}721 \\ \hline \end{array}
$$

7. $947 - 325$ $86 - 80$ $727 - 617$ $4{,}259 - 259$ $6{,}947 - 6{,}937$

D. Estimate the answer. Then subtract to find the exact answer.

8. In 1994, Sue realized it had been 23 years since she had been in high school. What was the last year she attended high school?

9. The computer package at Buy For Less is $1,997. If Comp-U-Town can beat that price by $423, what is the price there?

10. Felipe saved $7,255 to use as a down payment on a condo. If the condo costs $47,995, how much will he have to borrow?

11. **Explain** Last week Max drove 565 miles on his delivery route. How much farther will he drive if his new route is 578 miles? Explain how you would recognize this as a subtraction problem.

Deciding to Add or Subtract

How do you decide which operation to use to solve a word problem?

Choosing the Operation

Example: Clara drove 13 miles to work. Later she drove 16 miles to meet a friend. How many miles has she driven so far?

1. *Identify* the question. Restate it in your own words.

2. *Choose* the operation. To combine numbers, add. To find a difference or compare numbers, subtract.

3. *Estimate* first to find an approximate answer. Estimates will vary.

4. *Solve* to find an exact answer.

Think of this strategy when solving addition and subtraction problems.

Answer the questions about each problem.

1. Sumio put $2,525 down on a new car. If the sticker price of the car is $13,869, how much more will he have to pay?
   a. What's the question?
   b. Is it addition or subtraction? Why?
   c. Estimate the answer.
   d. What's the exact answer? Is it sensible?

2. Denver's football team gained 9 yards on one play and lost 5 yards on the next play. How many yards has Denver gained from where it started?
   a. What's the question?
   b. Is it addition or subtraction? Why?
   c. Estimate the answer.
   d. What's the exact answer? Is it sensible?

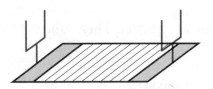

3. J.J. had a bank balance of $238. If he makes a deposit of $75, what will his new balance be?
   a. What's the question?
   b. Is it addition or subtraction? Why?
   c. Estimate the answer.
   d. What's the exact answer? Is it sensible?

Making Connections: Planning a Vacation

Let's plan a 10-day driving trip through Texas for 2 people. Look at the map for information to help solve the problems.

Day 1 begins in Dallas, with a short drive to Fort Worth to see the world famous Stockyards and visit Billy Bob's Country Western Saloon.

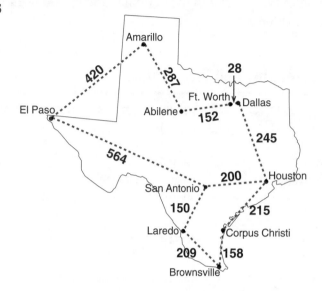

1. If expenses for the first day include $78 for a motel room, $42 for food, and $36 for sightseeing, what was the total bill for the day?

2. The next day is spent driving from Fort Worth to Amarillo by way of Abilene. How many more miles will you travel the second day than the first day?

3. Outside of Amarillo, you can enjoy horseback riding in the Palo Duro Canyon. The cost is $24 per person. That night you will see the majestic show *Texas* performed in the canyon. The price is $15 per person. It's an expensive day, but worth it. How much will you spend for 2 people for the day and evening?

4. Day 4 is a long drive to El Paso to see the Rio Grande River and the Mexican border. Day 5 is another long driving day to San Antonio. How many more miles was the drive on day 5 than on day 4?

5. On day 6 you can enjoy the beautiful old town of San Antonio and visit the Alamo.

In 1836 at the Alamo, 187 Texans rebelled against Mexican rule. All of them except 2 children and 2 women died. How many Texans died at the Alamo?

6. The drive continues the next day through Laredo and on to Brownsville on the Gulf of Mexico. Spend day 8 on the beach. What might be your expenses for this day? Make a list and find the total.

7. Day 9 is a long drive through Corpus Christi to Houston. How far have you driven from San Antonio to Houston? A short cut is shown on the map. How much shorter is the shortest way?

8. On the last day of the trip, get up early for a quick tour of the NASA space center in Houston. Later, begin the drive back to Dallas. How far have you traveled on the entire trip?

9. At one point, the United States measures 3,579 miles wide. How many miles more or less than this did you travel on this trip around Texas?

Subtracting by Regrouping

When you subtract, put the larger number on top. If a digit in the top number is smaller than a digit in the bottom number, you need to regroup (borrow) from a higher place value.

Regrouping and Place Value

Example 1: Joseph bought a shirt and a pair of pants on sale for $34. If the original price was $61, how much did he save?

Step 1	**Step 2**	**Step 3**	**Step 4**
Estimate first.	Put the larger number on the top. Line up the ones.	Subtract the ones.	Subtract the tens.
$60 − 30 ‾‾‾‾ $30	$61 − 34 ‾‾‾‾	$$\overset{5}{\cancel{6}}\,^{1}1$$ $6̸1$ − 34 ‾‾‾‾ 7	$$\overset{5}{\cancel{6}}\,^{1}1$$ $6̸1$ − 34 ‾‾‾‾ $27

⌐ 4 > 1
Regroup 1 ten.
10 + 1 = 11

Your answer of **$27** was close to your estimate of $30.

Example 2: $526 − $395

Step 1	**Step 2**	**Step 3**	**Step 4**	**Step 5**
Estimate first.	Line up the ones.	Subtract the ones.	Subtract the tens.	Subtract the hundreds.
$500 − 400 ‾‾‾‾ $100	$526 − 395 ‾‾‾‾	$526 − 395 ‾‾‾‾ 1	$$\overset{4}{\cancel{5}}\,^{1}26$$ $5̸26 − 395 ‾‾‾‾ 31	$$\overset{4}{\cancel{5}}\,^{1}26$$ $5̸26 − 395 ‾‾‾‾ $131

⌐ 9 > 2
Regroup 1 hundred.

Your answer of **$131** was close to your estimate of $100.

Example 3: 4,497 tickets − 2,545 tickets

Step 1	**Step 2**	**Step 3**	**Step 4**	**Step 5**
Estimate first.	Line up and subtract the ones.	Subtract the tens.	Subtract the hundreds.	Subtract the thousands.
4,500 − 2,500 ‾‾‾‾ 2,000	4,497 − 2,545 ‾‾‾‾ 2	4,497 − 2,545 ‾‾‾‾ 52	$$\overset{3}{\cancel{4}}\,^{1}497$$ 4̸,497 − 2,545 ‾‾‾‾ 952	$$\overset{3}{\cancel{4}}\,^{1}497$$ 4̸,497 − 2,545 ‾‾‾‾ 1,952

⌐ 5 > 4
Regroup 1 thousand.

Your answer of **1,952 tickets** was close to your estimate of 2,000.

A. Estimate first. Then practice regrouping as you subtract the given amounts.

1. 27   46   93   54   77
    − 9   − 38   − 49   − 8   − 58

2. 324   686   237   7,251   6,494
    − 53   − 94   − 145   − 340   − 873

B. For more practice, subtract, regrouping as necessary.

3. 63 − 8   925 − 16   90 − 28   65 − 27   281 − 90

4. 7,132 − 129   8,724 − 824   1,200 − 300   789 − 99

C. Use your subtraction skills to solve the problems below.

5. Explain how you could check this answer. Think of at least two ways.
   586 − 79 = 507

6. Subtract. Explain to a partner how you solved this problem.
   2,837 − 46

7. Write two subtraction equations based on 283 + 1,574 = 1,857. Discuss and
   explain these problems with a partner.

8. Write *true* or *false* after each problem.
   a. 549 − 58 > 500
   b. 9,237 − 2,427 < 7,000
   c. If 56 + 93 = 149, then 149 − 93 = 56.
   d. 3,000 − 900 < 2,000

9. **Estimate** Estimate the answer to this problem. Discuss your estimate with a
   partner. What is the exact answer?
   287 − 95

Mixed Review

A. Complete the subtraction problems.

1.
$$380 - 40$$
$$726 - 25$$
$$9{,}281 - 150$$
$$7{,}675 - 3{,}424$$

2. $9{,}298 - 285$ $584 - 13$ $777 - 647$

3.
$$92 - 43$$
$$725 - 18$$
$$427 - 318$$
$$6{,}721 - 820$$

4.
$$843 - 252$$
$$7{,}398 - 865$$
$$249 - 57$$
$$3{,}765 - 655$$

5. $872 - 9$ $4{,}366 - 274$ $6{,}845 - 2{,}934$

B. Subtract the amounts shown.

6.
$$894 - 384$$
$$2{,}514 - 593$$
$$64{,}239 - 3{,}818$$
$$78{,}423 - 6{,}503$$

7. $8{,}245 - 8{,}165$ $56{,}240 - 39$ $766 - 67$

8. $32{,}989 - 1{,}899$ $17{,}481 - 17{,}342$ $61{,}479 - 878$

C. Use your knowledge of subtraction to answer these problems.

9. Write two related subtraction facts for the addition fact $28 + 37 = 65$.

10. Write an addition fact related to the subtraction equation $294 - 75 = 219$.

11. Explain to a partner what happens when you subtract

    a. 0 from a number (Give an example.)

    b. 1 from a number (Give an example.)

    c. a number from itself (Give an example.)

12. Write a subtraction equation for each sentence below.

    a. Twelve less than forty-eight is thirty-six.

    b. The difference between sixty and nineteen is forty-one.

13. Find the value of the variable in each equation.

    a. $56 - 34 = N$ **b.** $39 - r = 12$ **c.** $A - 25 = 67$

       $N =$ $r =$ $A =$

D. You can use addition and subtraction to keep track of your bank accounts. When making a deposit, add. When making a withdrawal, subtract. Solve the problems below.

14. Elizabeth has $2,346 in her savings account. If she deposits $245, what is her new balance?

15. Tony and Linda saved $7,258 last year. If they withdraw $3,347 to buy a new computer and printer for their business, what is their new savings balance?

16. The Hayashi family saves money all year for their family vacation. Balance the bankbook below by filling in the balance on each line.

Hometown Savings Bank			
Date	Deposit	Withdrawal	Balance
1/1	$1,200		$1,200
3/1	$250		
5/1	$145		
5/20		$143	
7/1	$378		
8/1		$1,595	
8/20	$157		
10/1		$108	
11/1	$288		
12/30		$72	

Regrouping More than Once

Sometimes you have to regroup more than once. If so, keep track by crossing out and changing digits as necessary.

Regrouping to Subtract

Example 1: Marissa has $546 in her checking account. If she writes a check for $97, how much will she have left in the account?
Estimate: $550 − $100 = $450

Step 1	**Step 2**	**Step 3**	**Step 4**
Put the larger number on top. Line up the ones place.	Subtract the ones. Since 7 > 6, regroup 1 ten. $10 + 6 = 16$	Subtract the tens. Since 9 > 3, regroup 1 hundred. $100 = 10$ tens $10 + 3 = 13$	Subtract the hundreds.

$$
\begin{array}{r} 546 \\ -\ 97 \\ \hline \end{array}
\qquad
\begin{array}{r} \overset{3\ \ \ 1}{5\cancel{4}6} \\ -\ 97 \\ \hline 9 \end{array}
\qquad
\begin{array}{r} \overset{4\ 13\ \ 1}{\cancel{5}\cancel{4}6} \\ -\ 97 \\ \hline 49 \end{array}
\qquad
\begin{array}{r} \overset{4\ 13\ \ 1}{\cancel{5}\cancel{4}6} \\ -\ 97 \\ \hline 449 \end{array}
$$

Your answer of **$449** is close to your estimate of $450.

Example 2: 4,321 − 876
Explain the steps below to a partner.

Step 1	**Step 2**	**Step 3**	**Step 4**	**Step 5**
Estimate first.				

$$
\begin{array}{r} 4{,}300 \\ -\ 900 \\ \hline 3{,}400 \end{array}
\quad
\begin{array}{r} \overset{1\ \ 1}{4{,}3\cancel{2}1} \\ -\ 876 \\ \hline 5 \end{array}
\quad
\begin{array}{r} \overset{2\ 11\ \ 1}{4{,}\cancel{3}\cancel{2}1} \\ -\ 876 \\ \hline 45 \end{array}
\quad
\begin{array}{r} \overset{3\ 12\ 11\ \ 1}{\cancel{4}{,}\cancel{3}\cancel{2}1} \\ -\ 876 \\ \hline 445 \end{array}
\quad
\begin{array}{r} \overset{3\ 12\ 11\ \ 1}{\cancel{4}{,}\cancel{3}\cancel{2}1} \\ -\ 876 \\ \hline 3{,}445 \end{array}
$$

Your answer of **3,445** is close to your estimate of 3,400.

A. Estimate first. Then practice regrouping in these problems.

1.
$$
\begin{array}{r} 384 \\ -\ 96 \\ \hline \end{array}
\qquad
\begin{array}{r} 175 \\ -\ 88 \\ \hline \end{array}
\qquad
\begin{array}{r} 6{,}421 \\ -\ 554 \\ \hline \end{array}
\qquad
\begin{array}{r} 12{,}480 \\ -\ 3{,}682 \\ \hline \end{array}
$$

2. 352 − 63   4,384 − 575   68,222 − 9,153

B. For more practice, subtract these numbers.

3. 9,238 − 571   95,447 − 6,957   3,860 − 957   10,353 − 4,278

C. Use subtraction skills to solve this problem.

4. After the holidays, PCS Electronics had a stock reduction sale. The price list below shows the original and sale prices. In the savings column, write the amount saved on each item.

Item	Original	Sale	Savings
19" Color TV with Remote	$349	$299	_____
13" Color TV	$189	$129	_____
AM/FM Cassette Player	$195	$158	_____
Projection TV	$1,788	$1,499	_____
VCR	$229	$189	_____

Making Connections: Orbiting the Sun

The planets in our solar system all orbit (circle) the Sun. The chart below tells each planet's distance from the Sun and the number of days it takes to orbit the Sun. Use the chart to answer the questions.

Planet	Days of Orbit	Miles from the Sun
Mercury	88	36 million
Venus	225	67 million
Earth	365	93 million
Mars	687	141 million
Jupiter	4,380	484 million
Saturn	10,585	887 million
Uranus	30,660	1,780 million
Neptune	60,225	2,794 million
Pluto	90,520	3,658 million

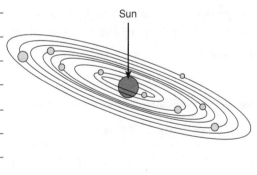

1. Which planet is closest to the Sun?

2. How much farther is Pluto than Saturn from the Sun?

3. a. Do you think the weather on Uranus will be hotter or colder than the weather on Earth? Why?

   b. What's the difference between the two planets' distance from the Sun?

4. How many more days does it take Neptune to orbit the Sun than it takes Venus?

5. How many more days does it take Pluto to orbit the Sun than it takes Earth?

6. Why do you think it takes so long for Pluto to orbit the Sun?

What Do I Need to Find?

When you read a word problem, let the words and ideas help you solve the problem.

- Read the problem carefully.
- Identify the question. Underline it.
- Look for certain common words that can help you understand what you need to find. Some common words are shown below.

Addition		Subtraction	
combine	sum	difference	remains
total	increase	left	how much more or less
altogether	and	change	decrease
plus			

Be careful when looking for these words. Some words can call for the opposite operation.

Example: Susan moved to a new apartment to save money. Her new rent is $200 a month, which is a *decrease* of $50 from her old rent. How much was her old rent?

Even though *decrease* is used, this is an addition problem.

$200 + $50 = **$250**

Use your real-life experiences to determine what the question is asking. What makes sense from the facts of the problem? Is the question asking you to combine or find the difference?

Identifying the Question

Example: Every week Andrew totals his sales for the week. Using the information below, underline find his sales for week 2.

Month: February

	Week 1	Week 2	Week 3	Week 4
Monday	$485	$795	$590	$565
Tuesday	$243	$200	$340	$325
Wednesday	$690	$485	$165	$180
Thursday	$511	$650	$748	$592
Friday	$384	$325	$295	$340

What do you need to find? Find the *total by adding* sales in week 2.

$795 + $200 + $485 + $650 + $325 = **$2,455**

Finding the Answer

Example: Andrew likes to compare his sales from week to week. <u>How much more were his sales on Thursday in week 3 than on Thursday in week 4?</u>

What do you need to find? Find the difference in sales between Thursday of week 4 and week 3 by subtracting. $748 − $592 = **$156**

A. Underline the question in each problem. Explain what you need to find to solve each problem. Then solve.

Fond du Lac	62
Oshkosh	81
Appleton	101
Green Bay	110

1. On her trip, Essie saw this road sign. When she got to Fond du Lac, how many more miles did she have to travel to get to Appleton?

ENROLLMENT
HARPER COMMUNITY COLLEGE

Year	Men	Women	Total
1992	12,148	11,807	23,955
1993	12,499	12,031	
1994	11,943		24,572

2. **a.** The enrollment figures for Harper Community College are shown on the right. What was the total number of students in 1993?

   b. How many women attended Harper in 1994?

Week	Pounds Lost
1	5
2	3
3	3
4	0
5	1

3. **a.** Rafael charted his progress on his diet. How much weight has he lost so far?

   b. If he wants to lose 35 pounds, how much weight does he have left to lose?

B. Betty and Ben are collectors. Betty has 356 antique buttons valued at about $1,800. Ben collects antique fishing lures. The value of his 97 lures is about $930. Solve these problems.

4. What is the combined value of the two antique collections?

5. How much more valuable is the antique button collection than the antique lures?

6. If Betty sells one of the antique buttons for $145, what is the value of her collection without that button?

7. **Multiple Solutions** Ben wants to increase his collection to 130 lures. Write two different equations to show how many lures Ben is thinking of adding. Then solve.

Subtracting from Zeros

When you need to regroup from a place value that contains a zero, move to the next column that contains a nonzero digit.

Regrouping and Zeros

Example: A concert hall that seats 8,024 has 163 scattered seats left. How many tickets for seats are sold?   8,024 − 163

Step 1	**Step 2**	**Step 3**	**Step 4**
Estimate first.	Subtract the ones.	Subtract the tens. Since 6 > 2, regroup 1 hundred from 80 hundreds.	Subtract the hundreds and the thousands.

Step 1
$$
\begin{array}{r}
8{,}000 \\
-\ 200 \\
\hline
7{,}800
\end{array}
$$

Step 2
$$
\begin{array}{r}
8{,}024 \\
-\ 163 \\
\hline
1
\end{array}
$$

Step 3
$$
\begin{array}{r}
{}^{7}\ {}^{9}\!\!{}^{1} \\
8{,}\cancel{0}2\,4 \\
-\ \ 163 \\
\hline
6\,1
\end{array}
$$

Step 4
$$
\begin{array}{r}
{}^{7}\ {}^{9}\!\!{}^{1} \\
8{,}\cancel{0}2\,4 \\
-\ \ 163 \\
\hline
7{,}8\,6\,1
\end{array}
$$

Your answer of **7,861 tickets** is close to your estimate of 7,800.

Example: 4,003 − 356

Step 1	**Step 2**	**Step 3**
Estimate first.	Subtract the ones. Since 6 > 3, regroup 1 ten from 400 tens.	Subtract the tens, hundreds, and thousands.

Step 1
$$
\begin{array}{r}
4{,}000 \\
-\ 400 \\
\hline
3{,}600
\end{array}
$$

Step 2
$$
\begin{array}{r}
{}^{3}\ {}^{9}\ {}^{9}\!\!{}^{1} \\
\cancel{4}{,}\cancel{0}\cancel{0}3 \\
-\ \ 356 \\
\hline
7
\end{array}
$$

Step 3
$$
\begin{array}{r}
{}^{3}\ {}^{9}\ {}^{9}\!\!{}^{1} \\
\cancel{4}{,}\cancel{0}\cancel{0}3 \\
-\ \ 356 \\
\hline
3{,}6\,4\,7
\end{array}
$$

Your answer of **3,647** was close to your estimate of 3,600.

A. Practice subtracting and regrouping on these problems.

1.
$$
\begin{array}{r} 502 \\ -\ 38 \\ \hline \end{array}
\qquad
\begin{array}{r} 3{,}400 \\ -\ 281 \\ \hline \end{array}
\qquad
\begin{array}{r} 900 \\ -\ 43 \\ \hline \end{array}
\qquad
\begin{array}{r} 6{,}000 \\ -\ 596 \\ \hline \end{array}
$$

2. $2{,}500 - 1{,}307$ $16{,}000 - 9{,}875$ $2{,}008 - 19$ $800 - 372$

B. For more practice, subtract these problems.

3. $200 - 145$ $6{,}007 - 598$ $39{,}000 - 25{,}640$

C. Practice your subtraction skills with this cross number puzzle. (*Hint:* First fill in the answers you can do quickly; see 36 down.)

Across

1. 5,943 – 2,482
5. 884 – 102
8. 9,403 – 3,179
9. 9,548 – 2,805
11. 7,243 – 3,153
13. 510 – 11
15. 700 – 26
17. 902 – 451
19. 84 – 19
20. 391 – 378
21. 66 – 42
22. 90 – 9
24. 135 – 38

26. 40 – 12
28. 6,000 – 5,302
30. 255 – 134
31. 800 – 95
33. 42,598 – 41,577
35. 7,020 – 4,010
37. 5,974 – 1,385
40. 492 – 69
41. 10,079 – 6,220

Down

1. 48 – 12
2. 143 – 101
3. 8,040 – 1,796
4. 560 – 420
5. 80,000 – 3,948
6. 100 – 13
7. 777 – 533
10. 436 – 40
12. 100 – 6
14. 150 – 55
15. 386 – 325
16. 864 – 791
18. 98 – 84

22. 112 – 26
23. 19,485 – 382
24. 102 – 10
25. 107 – 36
26. 68 – 41
27. 875 – 72
29. 120 – 40
30. 5,569 – 4,411
32. 1,000 – 496
34. 456 – 213
36. 15 – 3
38. 100 – 15
39. 400 – 301

Subtracting Dollars and Cents

The rules for subtracting money are very similar to those for adding money. Follow these tips when you subtract dollars and cents.

Subtracting Money

Tip 1. Line up place values when you subtract dollars.

Example: $356 – $45

$356
– 45
$311

Tip 2. Line up decimal points when you subtract dollars and cents. Subtract and regroup as if the decimal points were not there.

Example: $25 – $14.55
(Remember $25 is $25.00.)

$25.00
– 14.55
$10.45

Tip 3. When you subtract cents from dollars and cents, use the $ symbol and decimal point to show cents.

Example: $4.95 – 9¢ (Change 9¢ to $0.09.)

$4.95
– 0.09
$4.86

Tip 4. When you subtract money and the answer is less than ten cents, bring down the decimal point and put a zero in the dimes place.

Example: $1.96 – $1.94

$1.96
– 1.94
$.02

A. Subtract the amounts below.

1. $75 – $21 $250 – $48   56¢ – 48¢

2. $6,287 – $429 $3.25 – 8¢ $15.65 – 98¢

3. $37.48 – $37.45 $100 – $58.76 $3 – $1.84

B. For more practice, subtract the following money amounts.

4. $78 – $19 $26.50 – $7.75 $1.50 – $1.41

5. $16 – $7.99 $1 – 63¢ $10 – $6.80

6. $75,500 – $7,550 $20 – $12.88 $4,000 – $3,275

C. Solve the following.

7. Marta bought a $248 winter coat by paying $150 and putting the coat on layaway. How much does she have left to pay when she picks up the coat?

8. When Pedro drove through the toll booth, he gave the attendant a $10 bill to pay the 40¢ toll. What change should he receive?

9. Greg and Pat received an emergency room bill for $1,150. If their insurance paid $965.75, how much money do they need to pay the rest of the bill?

10. When Meredith bought a washer for $275.63, she charged it. When the bill came, she made a partial payment of $75. What is her new account balance before interest is added?

Making Connections: Tracking Vacation Spending

John and Janice decided to keep a record of their spending on their weeklong vacation in the Northwoods of Wisconsin. They began with $1,500. Find the balance after each expenditure.

Date	Description	Expenses	Balance
8/5	Beginning balance		$1,500.00
8/5	Gas for car	$15.40	_____
8/5 – 8/12	Lodging	$450.00	_____
8/5	Fishing license	$27.50	_____
8/6	Bait	$9.65	_____
8/6 – 8/12	Boat rental	$182.00	_____
8/6 – 8/12	Boat gasoline	$23.72	_____
8/12	Souvenirs	$86.35	_____
8/5 – 8/12	Food	$280.00	_____
8/12	Gas for car	$14.88	_____
8/13	Ending balance		_____

Understanding Your Paycheck

When you get your paycheck, you also receive a check stub that explains how your earnings have been distributed.

Custer Construction Company 4242 Mockingbird Lane Westmont, VA 24560	330-46-2706	2/11
	EMPLOYEE ID #	PAY DATE

PAY TO THE ORDER OF   Joanne May   529.77

Five hundred twenty-nine & $^{77}/100$ ————————————————— DOLLARS

PAYABLE THROUGH: First National
222 East Highland
Westmont, VA 24560

Jeff Foley

Look at the check stub below as you study the check information.

- **Gross pay** is your total earnings.
- **Net pay** is your take-home pay.
- **Deductions** are subtracted from your earnings by your employer.
- Net pay = Gross pay – Deductions
- **Tax deductions** include federal and state taxes.
- **FICA** (Federal Insurance Contributions Act) is the Social Security deduction.
- **Insurance deductions** can be medical or life insurance payments.
- **Special deductions** may include union dues and charitable contributions.
- Notice each current amount is added to the year-to-date totals.
- Personal information includes name, Social Security number, rate of pay, exemptions, and hours worked. Always check your personal information to make sure it is correct.

CUSTER CONSTRUCTION COMPANY			NO. 116583	
EMPLOYEE NAME: Joanne May SOCIAL SECURITY #: 330-46-2706			EXEMPTIONS: Fed 1 State 1 Status 1 PAY PERIOD: 1/14 to 1/28	
CURRENT EARNINGS			DEDUCTIONS: Current/Year-to-Date	
	Hours/Rate/Pay		FEDERAL	136.17	258.57
REGULAR	80 8.50	680.00	STATE	15.13	28.56
OVERTIME	6 12.75	76.50	FICA	52.96	100.56
PAY	Current/Year-to-Date		INSURANCE	7.57	14.37
GROSS	756.50 1436.50		UNION	12.40	24.80
NET	529.77 1004.64		UNITED WAY	2.50	5.00
VACATION BALANCE: 10 DAYS			TOTAL 226.73 431.86	

A. Use the sample check stub at the bottom of page 92 and the check information tips above it to answer the following questions.

1. How much did Joanne earn on this current check?

2. How much did she take home?

3. Find the difference between what she earned and the amount she took home.

4. How much more does Joanne earn per overtime hour than per regular hour?

5. Write an equation that describes how you calculate the current net pay.

6. How many total hours did Joanne work during this pay period?

7. Write an equation to describe how you calculate Joanne's current gross pay from the information given.

8. If Joanne did not belong to a union and made no charitable contributions, how much would her current total deductions be?

B. Work with a partner to check your understanding of paycheck stubs. Fill in the missing information on Joanne's next paycheck stub shown below. Use information from the check stub on the previous page when necessary.

CUSTER CONSTRUCTION COMPANY			NO. 116735	
EMPLOYEE NAME: Joanne May SOCIAL SECURITY #: 330-46-2706		**EXEMPTIONS:** Fed 1 State 1 Status 1 PAY PERIOD: 1/29 to 2/11		
CURRENT EARNINGS		**DEDUCTIONS:** Current/Year-to-Date		
Hours/Rate/Pay		FEDERAL	143.06	401.63
REGULAR 80 8.50 680.00		STATE	15.90	44.46
OVERTIME 9 12.75 114.75		FICA	55.63	156.19
PAY Current/Year-to-Date		INSURANCE	15.90	30.27
GROSS ____ ____		UNION	12.40	37.20
NET ____ ____		UNITED WAY	2.50	7.50
VACATION BALANCE: 10 DAYS		TOTAL	____	____

Subtracting on a Calculator

When using a calculator, always check to make sure the displayed answer makes sense. It is helpful to estimate your answer when you use a calculator in case you enter a number incorrectly.

Subtracting Using a Calculator

Example: 725 – 36

	Key in:	Your display reads:	
Press On/Clear.	On/C	0.	Estimate.
Enter the largest number.	7 2 5	725.	730
Press the minus key.	–	725.	– 40
Enter the amount being subtracted.	3 6	36.	690
Press the equal sign.	=	689. ← difference	

Practice this procedure on your calculator.

Points to remember when you subtract on a calculator:

- Order is important when you subtract. Be sure the number being subtracted is entered *after* the minus sign.

   Example: Nine less than twenty is 20 – 9.

- When you subtract money amounts, you cannot use the $ symbol. Enter the dollar digits, a decimal point, and the cents digits.

- Enter cents only with a decimal and two digits.
   Example: Enter 48¢ as · 4 8 .

- When you enter dollars and cents, zero digits after the decimal point are not displayed if they are the last digits.

   Example: $1.50 will be displayed as 1.5
   $16.00 will be displayed as 16.
   $.30 will be displayed as 0.3

- If the answer to a subtraction money problem has only one digit after the decimal point, attach a zero. The number of cents must always have two digits.

   Example: $1.98 – $1.28
   You can write the answer display reading 0.7 as **$0.70** *or* **$.70.**

A. Use a calculator to solve the problems below.

1. 389 − 296   4,726 − 3,949   76,003 − 29,157   248 − 97

2. $7.86 − $3.84 $5 − 36¢ $56 − $14.65 $725.37 − $98.17

B. For more practice, write a subtraction equation for each expression and estimate an answer. Use the calculator to find the exact answer.

3. Find the number that is 67 less than 524.

   _____ − _____ = n

   Estimate: _____   Exact: n = _____

4. What is 3,472 take away 2,905?

   _____ − _____ = W

   Estimate: _____   Exact: W = _____

5. Find the difference between 604 and 575.

   _____ − _____ = d

   Estimate: _____   Exact: d = _____

6. If you take $485.50 away from $733.60, how much is left?

   _____ − _____ = L

   Estimate: _____   Exact: L = _____

You may subtract two or more numbers on a calculator. Press the minus sign before each number being subtracted. Press the equal sign after you enter the final number.

C. Estimate an answer and then use your calculator to get an exact answer.

7. Josh had $25 in his pocket when he stopped at the grocery store to pick up a few items. He used a calculator as he shopped to make sure he didn't spend too much. He entered $25 and began subtracting as he put each item in his cart. His purchases are shown below. Find the amount he had left when he finished shopping.

   $2.79 gallon of milk
   $6.45 beef roast
   $1.98 BBQ sauce
   $3.99 5 lb. oranges
   $2.48 jar of peanut butter
   $1.99 vegetables

Figuring Change

A cashier first enters the amount you paid into the cash register. Then the cashier subtracts the cost of the purchase. Whether you are the cashier or the customer, it is important to be able to enter and calculate the correct change.

Use subtraction to calculate the amount of change.

Amount paid − Total cost = Change

Example: You pay: $25
Total cost: $21.80

$25.00
− 21.80
$3.20 ←——— your change

For another look
at money, turn to
page 206.

 A. Find the amount of change you should receive if you pay for each purchase below with a $20 bill. You may use a calculator.

1. lunch for two
$14.72

school supplies
$8.95

bouquet of flowers
$17.45

2. gallon of milk
$2.79

movie ticket
$6.50

cosmetics
$11.06

 B. Use subtraction to find the amount of change that is due in each situation below. You may use a calculator.

3. You pay $5 for a $3.25 purchase. How much is your change?

4. Dan bought groceries for $87.25 and paid with a check for $122.25. How much extra cash did he get back?

5. How much change did Shaheena receive when she paid for a 68¢ candy bar with $1?

6. At the restaurant, Judy paid $40 for a $34.50 bill. She told the waiter to keep the change. How much was the waiter's tip?

Making Change

If you work as a cashier, you will have to select the bills and coins that total the amount of change.

Example: A customer buys athletic shoes for $48.72 and pays $60.

$$\begin{array}{r} \$60.00 \\ -\ 48.72 \\ \hline \$11.28 \end{array}$$

The change is: **$11.28.**

The bills and coins are: $10 bill, $1 bill, 25¢, and 1¢, 1¢, 1¢.

 C. Figure the change and select the bills and coins you would use to make the change. All prices listed include tax. You may use a calculator.

7. 4 tires: $125.24

Customer pays: $140.00

a. The change is:

b. The bills and coins are:

8. Jeans: $29.11

Customer pays: $30

a. The change is:

b. The bills and coins are:

9. Phone/answering machine: $75.59

Customer pays: $100

a. The change is:

b. The bills and coins are:

10. Video: $21.55

Customer pays: $40

a. The change is:

b. The bills and coins are:

D. Multiple Answers Take a shopping spree. You have a $100 bill. Buy three items from the choices below. Estimate to find the total cost. Then figure the approximate change you'll receive when you pay with the $100 bill. What bills will you receive?

Comforter, any size
$29.99

Pillows
$8.99 each

Bath scale
$39.99

Table cloth
$9.99

Towels
$4.99 each

Twin sheet sets
$14.99

Item 1: _____

Item 2: _____

Item 3: _____

Subtotal: _____

Tax: _____2.87_____

Total: _____

You pay: $100

Your change is: _____

The bills are: _____

Unit 3 Review

A. Write a number in each blank to make each statement true. Some of these questions may have more than one correct answer.

1. _____ is three less than ten.

2. _____ is a three-digit number.

3. _____ is an even number.

4. _____ is an odd number between 5 and 9.

5. In 24,386, _____ is in the hundreds place.

6. _____ < 17

7. One foot = _____ inches

8. _____ > 4,357

9. $369 rounded to the nearest $100 is _____.

10. One year is about _____ days.

11. _____ is half past nine in the morning.

12. 56 + _____ = 56

13. 8 + 2 = _____ + 6

14. 7 + _____ + 9 = 19

15. $A + A = 18, A = $ _____

16. 43 + 9 = _____

17. 390 + 65 + 7 = _____

18. 9,253 + 7,084 = _____

19. The perimeter of a rectangle with a 9-foot width and 12-foot length is _____.

20. 48¢ + 87¢ = _____

B. Solve the problems below.

21. **a.**

17	38	14	16
− 0	− 38	− 9	− 8

b.

12	19	15	13
− 7	− 8	− 7	− 5

22.

472	98	2,044	686	7,430
− 351	− 38	− 1,040	− 99	− 941

23. 952 − 915   671 − 85   3,294 − 2,585

24. 800 − 247   5,006 − 3,877   37,000 − 18,901

25. $t − 9 = 12$ $18 − n = 7$ $34 − 19 = D$

    $t =$ $n =$ $D =$

C. Write an equation to help you solve each problem. Then solve.

26. In 1994 Jeremy's Ski School celebrated its 25th anniversary. What year did the school open?

27. On Monday Sean had a balance of $2,846 in his savings account. If he deposits $350 on Wednesday and $189 on Thursday and then withdraws $296 on Saturday, what is his new balance?

28. U.S. Motors has a $750 rebate program for first-time buyers. If you buy a $12,600 car, how much will it cost after the rebate?

29. How much change will you receive if you make a $47.38 purchase and pay with three $20 bills? What bills and coins will you receive?

30. When Jocko bought a new car, he was disappointed with what the dealer gave him for his trade-in. The new car sticker price was $15,500. After the trade-in, he paid $13,875. How much did he get for his trade-in?

Working Together

Learning to communicate with others when working with math will help you develop your own problem-solving skills.

Work with a partner to discuss and investigate the following problems.

1. What have you been doing recently to increase your confidence in your ability to do math?

2. Discuss how addition and subtraction problems are alike and different.

3. Explain how to subtract 478 from 9,000.

4. Write four related addition and subtraction facts using the digits 9, 6, and 3.

5. Write four related addition and subtraction facts using another group of digits of your own choosing.

Unit 4

Multiplication

Multiplication is a shortcut to addition when all the numbers are the same. For example, you could use multiplication when you

- figure the amount earned at work
- estimate the total cost of taking your family to the movies

Suppose you earn $8 per hour and work 6 hours. You could add $8 + $8 + $8 + $8 + $8 + $8 = $48. Or, because you are paid the same each hour, you could use the shortcut $8 × 6 = $48. The **times sign** × shows multiplication.

Words that may indicate multiplication include *times, total, product, of, in all, twice, altogether, double,* and *triple.*

First you must learn the multiplication facts. After mastering these facts, you will use them in problem solving.

100

When Do I Multiply?

Which skill would you use in each situation below? Write *A* for addition, *S* for subtraction, and *M* for multiplication.

_____ **1.** I gave my daughter $3 and my son $6. What is the total?

_____ **2.** I paid car payments of $145 per month for a year. What was my total payment for the year?

_____ **3.** I bought 5 gift certificates at $15 each for my friends. How much did I spend?

_____ **4.** If I cut 5 inches off the 36-inch board to make my shelf fit, how long is the board that remains?

_____ **5.** When I traveled across the state, I drove 55 miles per hour for 4 hours. How far did I drive?

6. Remember that the numbers have to be the same when you use multiplication as a shortcut for addition. Rewrite the problems below as multiplication problems if it's possible.

a.	b.	c.
3	8	$9
3	6	+ $7
3	7	
+ 3	+ 9	

d. $56
+ $56

e. 6 + 6 + 6 + 6 =

7. Counting patterns can help you learn the multiplication facts.

Count by twos:

2   4   6   8   10   ___   ___   ___   18   20

Count by fives:

5   10   15   ___   ___   30   ___   ___   ___   50

Make up your own pattern to count by.

When learning the multiplication facts, you may want to pay special attention to these six most often missed facts:

$9 \times 8 = 72$	$8 \times 7 = 56$
$9 \times 7 = 63$	$8 \times 6 = 48$
$9 \times 6 = 54$	$7 \times 6 = 42$

Talk About It

Discuss which of the ways listed below would help you learn the six most often missed multiplication facts. Place a check mark next to the ones you choose.

☐ Memorize one fact each day.

☐ Use flash cards by myself or with a friend.

☐ Say the problems repeatedly several times a day.

☐ Write the problems on paper when I have a break.

☐ Think of a real-life problem when I could use these numbers. (Example: On vacation I bought 7 souvenirs at $6 each. $7 \times \$6 = \$42$)

Building a Multiplication Table

You can rely on your counting skills to set up a **multiplication table.** For example, as you count by threes, the numbers become the multiples of 3: 3, 6, 9, 12, 15, etc. A **multiple** of 3 is a product of 3 and another number.

A. You can build a multiplication table by following the steps below. Steps 1–4 have been done for you. Complete the table by following step 5.

Step	Reason
1. List all the numbers 0 to 10 across the top row.	**1.** This row will provide one of the numbers to multiply by.
2. List all the numbers 0 to 10 down the first column.	**2.** This column will provide the other number to multiply by.
3. Write 0 across the second row and down the second column.	**3.** When you multiply by 0, the answer is always 0.
4. Copy the first row into the third row and the first column into the third column.	**4.** When you multiply by 1, the answer is always the original number.
5. Fill in the rest of the chart by counting by the first number in the column. (For example, count by twos: 0, 2, 4, 6, 8, 10.)	**5.** Multiplication is a shortcut for repeated addition. (Remember, 2 × 3 is the same as 2 + 2 + 2.)

Multiplication Table

×	0	1	2	3	4	5	6	7	8	9	10
0	0	0	0	0	0	0	0	0	0	0	0
1	0	1	2	3	4	5	6	7	8	9	10
2	0	2			8						
3	0	3			12						
4	0	4			16						
5	0	5			20						
6	0	6			24						
7	0	7	14	21	28	35	42	49	56	63	70
8	0	8			32						
9	0	9			36						
10	0	10			40						

Now you can use the table to discover multiplication facts.

Example: What is 7 × 4? Find where the row for 7 and the column for 4 meet. The answer is **28.**

B. Use the multiplication table to answer these questions.

1. The table shows four ways to get the answer of 24. What are they?

____ × ____ = 24   ____ × ____ = 24   ____ × ____ = 24   ____ × ____ = 24

2. $3 \times 0 =$ ____
   $7 \times$ ____ $= 0$
   ____ $\times 0 = 0$
   $2 \times$ ____ $= 0$

3. $6 \times 1 =$ ____
   ____ $\times 1 = 9$
   $4 \times 1 =$ ____
   ____ $\times 1 = 5$
   $8 \times$ ____ $= 8$

4. $2 \times 0 =$ ____
   $2 \times 1 =$ ____
   $2 \times$ ____ $= 4$
   ____ $\times 3 = 6$
   $2 \times 4 =$ ____
   $2 \times$ ____ $= 10$
   ____ $\times 6 = 12$
   $2 \times 7 =$ ____
   ____ $\times 8 = 16$
   $2 \times$ ____ $= 18$
   $2 \times 10 =$ ____

5. $3 \times$ ____ $= 0$
   ____ $\times 1 = 3$
   $3 \times 2 =$ ____
   $3 \times$ ____ $= 9$
   ____ $\times 4 = 12$
   $3 \times 5 =$ ____
   ____ $\times 6 = 18$
   $3 \times 7 =$ ____
   $3 \times$ ____ $= 24$
   ____ $\times 9 = 27$
   $3 \times 10 =$ ____

6. $4 \times$ ____ $= 4$
   $4 \times 2 =$ ____
   $4 \times 3 =$ ____
   ____ $\times 4 = 16$
   $4 \times$ ____ $= 20$
   $4 \times 6 =$ ____
   $4 \times$ ____ $= 28$
   $4 \times 8 =$ ____
   $4 \times 9 =$ ____
   $4 \times$ ____ $= 40$

7. $5 \times$ ____ $= 0$
   $5 \times 1 =$ ____
   $5 \times$ ____ $= 10$
   ____ $\times 3 = 15$
   $5 \times 4 =$ ____
   $5 \times 5 =$ ____
   ____ $\times 6 = 30$
   $5 \times$ ____ $= 35$
   $5 \times$ ____ $= 40$
   $5 \times 9 =$ ____
   $5 \times 10 =$ ____

8. $6 \times$ ____ $= 6$
   $6 \times 2 =$ ____
   $6 \times 3 =$ ____
   $6 \times$ ____ $= 24$
   $6 \times 5 =$ ____
   $6 \times$ ____ $= 36$
   $6 \times 7 =$ ____
   $6 \times 8 =$ ____
   $6 \times$ ____ $= 54$
   $6 \times$ ____ $= 60$

C. Use your counting skills and the table to complete these facts.

9. $7 \times$ ____ $= 7$
   $7 \times 2 =$ ____
   ____ $\times 3 = 21$
   $7 \times 4 =$ ____
   $7 \times 5 =$ ____
   $7 \times$ ____ $= 42$
   $7 \times 7 =$ ____
   $7 \times 8 =$ ____
   $7 \times$ ____ $= 63$

10. $8 \times$ ____ $= 16$
    $8 \times 3 =$ ____
    $8 \times 4 =$ ____
    $8 \times$ ____ $= 40$
    ____ $\times 6 = 48$
    $8 \times 7 =$ ____
    $8 \times 8 =$ ____
    $8 \times$ ____ $= 72$
    ____ $\times 10 = 80$

11. $9 \times$ ____ $= 18$
    $9 \times 3 =$ ____
    $9 \times$ ____ $= 36$
    $9 \times$ ____ $= 45$
    $9 \times 6 =$ ____
    $9 \times 7 =$ ____
    $9 \times$ ____ $= 72$
    ____ $\times$ ____ $= 81$
    $9 \times$ ____ $= 90$

Multiplication Strategies

You can show multiplication on a number line. To picture 2×4, begin at 0 and mark off 2 units 4 times. $2 \times 4 = 8$

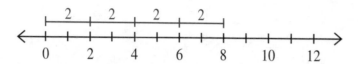

A. Use the number lines to fill in the blanks. Look for patterns.

1.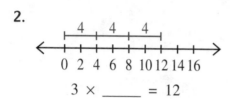

    $5 \times \underline{\hspace{1cm}} = 10$

4.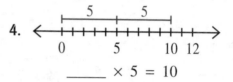

    $\underline{\hspace{1cm}} \times 5 = 10$

2.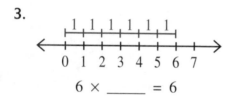

    $3 \times \underline{\hspace{1cm}} = 12$

5.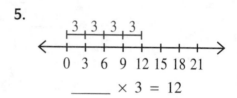

    $\underline{\hspace{1cm}} \times 3 = 12$

3.
    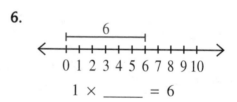

    $6 \times \underline{\hspace{1cm}} = 6$

6.

    $1 \times \underline{\hspace{1cm}} = 6$

The number line and the multiplication table are tools you can use to memorize the multiplication facts. Practicing counting by sixes, sevens, eights, and nines will also help you learn the more difficult facts. By memorizing these facts, you will gain confidence in your mathematical ability and have a solid foundation for future problem solving.

Take time now to memorize the multiplication facts.

Here are some basics you should be familiar with:

- The numbers being multiplied are called **factors.** The answer is the **product.** The number doing the multiplying is also called the **multiplier.**

- Multiplication can be written two ways:

    horizontally   *or*   **vertically**

    $7 \times 8 = 56$

    $$\begin{array}{r} 8 \\ \times\ 7 \\ \hline 56 \end{array}$$

Patterns in Multiplication

Here are some tips to help you when multiplying:

- You can multiply *in any order.*

    Example: $7 \times 8 = 56$ and $8 \times 7 = 56$

- **Twos** The multiples of 2 are *even numbers.*

    Examples: 0, 2, 4, 6, 8, 10, 12, 14, 16, 18, 20

- **Fives** The last digit of the multiples of 5 are 5 or 0.

    Examples: 5, 10, 15, 20, 25, 30, 35, 40, 45, 50

- **Tens** The last digit of the multiples of 10 is 0.

    Examples: 10, 20, 30, 40, 50, 60, 70, 80, 90, 100

- **Nines** Look for patterns in the multiples of 9 shown below.

$9 \times 1 = 9$	As you write the multiples, each consecutive answer begins
$9 \times 2 = 18$	with the next counting number.
$9 \times 3 = 27$	Each answer begins with a digit that is one less than the multiplier.
$9 \times 4 = 36$	Examples: $9 \times 6 = 54$
$9 \times 5 = 45$	
$9 \times 6 = 54$	 5 is one less than 6.
$9 \times 7 = 63$	$9 \times 7 = 63$
$9 \times 8 = 72$	 6 is one less than 7.
$9 \times 9 = 81$	The sum of the digits in the product is 9.
	Example: For the answer 36, $3 + 6 = 9$.

B. Use the multiplication tips to answer the problems below.

7. $3 \times 6 = \underline{\hspace{1cm}} \times 3$ $\qquad$ $4 \times 2 = \underline{\hspace{1cm}}$ $\qquad$ $6 \times 5 = \underline{\hspace{1cm}}$

 $5 \times 2 = 2 \times \underline{\hspace{1cm}}$ $\qquad$ $7 \times 2 = \underline{\hspace{1cm}}$ $\qquad$ $5 \times 4 = \underline{\hspace{1cm}}$

8. $7 \times 6 = \underline{\hspace{1cm}}$ $\qquad$ $9 \times 6 = \underline{\hspace{1cm}}$ $\qquad$ $6 \times 10 = \underline{\hspace{1cm}}$

 $6 \times \underline{\hspace{1cm}} = 42$ $\qquad$ $\underline{\hspace{1cm}} \times 9 = 54$ $\qquad$ $\underline{\hspace{1cm}} \times 8 = 80$

C. Discuss the following questions with a partner. Give examples.

9. a. How do you recognize the multiples of 5?

 **b.** of 10?

10. There are special patterns in the multiples of 9. Describe them.

11. Do you remember these difficult facts?

$8 \times 6 =$ $\qquad\qquad$ $8 \times 7 =$ $\qquad\qquad$ $9 \times 8 =$

$9 \times 7 =$ $\qquad\qquad$ $7 \times 6 =$ $\qquad\qquad$ $9 \times 6 =$

Multiplication Facts

The multiplication facts are the basis of your work with multiplication and division.

A. Practice your multiplication facts on these problems. Practice these problems until you can do them quickly with no errors.

1. a.
$$\begin{array}{r} 6 \\ \times\ 2 \\ \hline \end{array} \qquad \begin{array}{r} 4 \\ \times\ 8 \\ \hline \end{array} \qquad \begin{array}{r} 9 \\ \times\ 7 \\ \hline \end{array}$$
b.
$$\begin{array}{r} 6 \\ \times\ 6 \\ \hline \end{array} \qquad \begin{array}{r} 3 \\ \times\ 4 \\ \hline \end{array} \qquad \begin{array}{r} 9 \\ \times\ 2 \\ \hline \end{array}$$

2. a.
$$\begin{array}{r} 7 \\ \times\ 0 \\ \hline \end{array} \qquad \begin{array}{r} 8 \\ \times\ 3 \\ \hline \end{array} \qquad \begin{array}{r} 2 \\ \times\ 7 \\ \hline \end{array}$$
b.
$$\begin{array}{r} 5 \\ \times\ 5 \\ \hline \end{array} \qquad \begin{array}{r} 3 \\ \times\ 0 \\ \hline \end{array} \qquad \begin{array}{r} 5 \\ \times\ 4 \\ \hline \end{array}$$

3. a. $6 \times 4 =$ $8 \times 8 =$ $9 \times 6 =$ b. $6 \times 7 =$ $8 \times 5 =$ $4 \times 2 =$

4. a. $4 \times 9 =$ $5 \times 3 =$ $7 \times 4 =$ b. $6 \times 8 =$ $8 \times 2 =$ $5 \times 9 =$

5. a. $5 \times 6 =$ $2 \times 5 =$ $9 \times 8 =$ b. $7 \times 5 =$ $3 \times 6 =$ $7 \times 9 =$

6. a.
$$\begin{array}{r} 7 \\ \times\ 9 \\ \hline \end{array} \qquad \begin{array}{r} 4 \\ \times\ 4 \\ \hline \end{array} \qquad \begin{array}{r} 8 \\ \times\ 6 \\ \hline \end{array}$$
b.
$$\begin{array}{r} 1 \\ \times\ 1 \\ \hline \end{array} \qquad \begin{array}{r} 6 \\ \times\ 1 \\ \hline \end{array} \qquad \begin{array}{r} 9 \\ \times\ 3 \\ \hline \end{array}$$

7. a.
$$\begin{array}{r} 2 \\ \times\ 5 \\ \hline \end{array} \qquad \begin{array}{r} 8 \\ \times\ 4 \\ \hline \end{array} \qquad \begin{array}{r} 3 \\ \times\ 5 \\ \hline \end{array}$$
b.
$$\begin{array}{r} 6 \\ \times\ 5 \\ \hline \end{array} \qquad \begin{array}{r} 0 \\ \times\ 6 \\ \hline \end{array} \qquad \begin{array}{r} 7 \\ \times\ 7 \\ \hline \end{array}$$

8. a. $5 \times 8 =$ $5 \times 7 =$ $3 \times 8 =$ b. $6 \times 9 =$ $6 \times 3 =$ $1 \times 9 =$

9. a. $9 \times 4 =$ $4 \times 5 =$ $2 \times 4 =$ b. $4 \times 7 =$ $7 \times 6 =$ $8 \times 6 =$

10. a. $9 \times 5 =$ $2 \times 6 =$ $5 \times 2 =$ b. $9 \times 0 =$ $8 \times 4 =$ $8 \times 7 =$

11. a.
$$\begin{array}{r} 8 \\ \times\ 9 \\ \hline \end{array} \qquad \begin{array}{r} 2 \\ \times\ 2 \\ \hline \end{array} \qquad \begin{array}{r} 1 \\ \times\ 7 \\ \hline \end{array}$$
b.
$$\begin{array}{r} 8 \\ \times\ 1 \\ \hline \end{array} \qquad \begin{array}{r} 4 \\ \times\ 3 \\ \hline \end{array} \qquad \begin{array}{r} 9 \\ \times\ 9 \\ \hline \end{array}$$

12. a.
$$\begin{array}{r} 0 \\ \times\ 4 \\ \hline \end{array} \qquad \begin{array}{r} 2 \\ \times\ 1 \\ \hline \end{array} \qquad \begin{array}{r} 7 \\ \times\ 8 \\ \hline \end{array}$$
b.
$$\begin{array}{r} 9 \\ \times\ 3 \\ \hline \end{array} \qquad \begin{array}{r} 3 \\ \times\ 3 \\ \hline \end{array} \qquad \begin{array}{r} 2 \\ \times\ 9 \\ \hline \end{array}$$

Multiplication Equations

A **multiplication equation** is a number sentence that combines equal amounts together. There is more than one way to show multiplication in an equation.

An equation can be written using a times sign $\times$ and an equal sign. Another way to write the equation is to use parentheses and an equal sign. The parentheses indicate multiplication.

Examples: $7 \times 8 = 56$ or $7(8) = 56$
$6 \times 9 = 54$ or $6(9) = 54$

B. Write the following sentences as multiplication equations.

13. Eight times five equals forty.

14. The product of three and eight is twenty-four.

15. Seven multiplied by itself is forty-nine.

16. Six times eight totals forty-eight.

If the multiplier in an equation is a variable, show multiplication by writing the number next to the variable.

Examples: $6n = 48$ (Six times a number n equals 48.)
$3A = 15$ (Three times a number A is 15.)

C. Find the value of the variable in each equation below.

17. $4(7) = T$

$T =$

$5n = 50$

$n =$

Hint: $5 \times \underline{\quad} = 50$

$7 \times 8 = W$

$W =$

18. $6y = 48$

$y =$

Hint: $6 \times \underline{\quad} = 48$

$9c = 54$

$c =$

$3(9) = G$

$G =$

D. Write an equation for each sentence below. Use a variable (letter) for the unknown number. The first one is started for you.

19. What number times 8 is 56?

$8n =$

21. What number is 6 times 7?

20. What number is 7 less than 12?

22. What number added to 8 is 13?

Multiplying by One-Digit Numbers

If a multiplication problem is given horizontally, you may want to rewrite it vertically to make it easier to work. Be sure to put the larger number on top. The lower number is the multiplier. First, multiply the ones place by the multiplier, then the next places in turn.

Multiplying

Example: Martin takes home $334 a paycheck. He gets paid twice a month. How much does he take home in a month?

$334 × 2   Estimate first: $350 × 2 = $700

Step 1	**Step 2**	**Step 3**	**Step 4**
Put the larger number on top.	Multiply the ones.	Multiply the tens.	Multiply the hundreds.
$334 × 2	$334 × 2 8	$334 × 2 68	$334 × 2 $668

There are two ways you might check your answer.

You could check against your estimate.   *or*   Multiply again and check if

$668 is close to your estimate of $700.

- your digits are lined up correctly
- you multiplied correctly
- your answer makes sense

A. Multiply the problems below. Check your answers.

1. a.
123   12   11   **b.** 40   110   201
× 3   × 4   × 6   × 2   × 7   × 4

2. a.
31   603   75   **b.** 80   57   92
× 5   × 2   × 1   × 7   × 0   × 3

B. For more practice, multiply these numbers.

3. 401 × 6   92 × 3   81 × 9   800 × 7

4. 420 × 2   8,111 × 6   7,000 × 7   510 × 5

Multiplying Three or More Numbers

When you multiply three or more numbers, multiply two at a time. Then multiply that product by the next number. *The numbers may be multiplied in any order.*

Example: $2 \times 5 \times 8$

$$
\begin{array}{cc}
2 & \longrightarrow 10 \\
\underline{\times\ 5} & \underline{\times\ 8} \\
10 \longrightarrow & \mathbf{80}
\end{array}
\qquad or \qquad
\begin{array}{cc}
5 & \longrightarrow 40 \\
\underline{\times\ 8} & \underline{\times\ 2} \\
40 \longrightarrow & \mathbf{80}
\end{array}
$$

C. Multiply the numbers below.

5. $5 \times 6 \times 8$ $3 \times 11 \times 3$ $4 \times 1 \times 22$ $6 \times 9 \times 0$

6. $12 \times 2 \times 2$ $3 \times 4 \times 4$ $8 \times 5 \times 5$ $3 \times 7 \times 4$

D. Use your multiplication skills to solve the following problems.

7. How much will you save at the Montana Motel if you pay the weekly rate instead of the daily rate for 7 days?

> **Montana Motel**
> **Daily Rate: $41 per person**
> **Weekly Rate: $250 per person**

8. Fill in the missing numbers.

×	20	91	501
7			3,507
3		273	
6	120		

9. Abraham Lincoln began a famous speech "Four score and seven years ago. . . ." If a score is 20, how many years ago was he talking about?

10. Estimate Reiko usually spends $73 per week on groceries. When her son's family visited for a week, the grocery bill tripled. *About* how much was her bill that week?

Multiplying and Regrouping

Like addition, multiplication often requires you to regroup. When the product is 10 or more, regroup to the next place. Be sure to multiply the next place before you add the regrouped amount.

Regrouping in Multiplication

Example: Jasmine sold 3 gift baskets for $27 each. Instead of adding, she can multiply to find the amount of money she took in. $27 × 3 = ?

Step 1	**Step 2**	**Step 3**	**Step 4**
Estimate first.	Put the larger number on top.	Multiply the ones. Regroup the 2.	Multiply the tens. Add the 2.

Step 1
$$\begin{array}{r} \$30 \\ \times\ 3 \\ \hline \$90 \end{array}$$

Step 2
$$\begin{array}{r} \$27 \\ \times\ 3 \\ \hline \end{array}$$

Step 3
$$\begin{array}{r} {}^{2}\ \\ \$27 \\ \times\ 3 \\ \hline 1 \end{array}$$

Step 4
$$\begin{array}{r} {}^{2}\ \\ \$27 \\ \times\ 3 \\ \hline 81 \end{array}$$

The answer **$81** is close to the estimate of $90.

A. Estimate, then find an exact answer. Regroup when necessary.

1. a.

$$\begin{array}{r} 86 \\ \times\ 3 \\ \hline \end{array} \qquad \begin{array}{r} 35 \\ \times\ 9 \\ \hline \end{array} \qquad \begin{array}{r} 47 \\ \times\ 2 \\ \hline \end{array}$$

b.

$$\begin{array}{r} 16 \\ \times\ 6 \\ \hline \end{array} \qquad \begin{array}{r} 59 \\ \times\ 8 \\ \hline \end{array} \qquad \begin{array}{r} 73 \\ \times\ 7 \\ \hline \end{array}$$

2. a.

$$\begin{array}{r} 752 \\ \times\ 4 \\ \hline \end{array} \qquad \begin{array}{r} 350 \\ \times\ 9 \\ \hline \end{array} \qquad \begin{array}{r} 603 \\ \times\ 7 \\ \hline \end{array}$$

b.

$$\begin{array}{r} 209 \\ \times\ 8 \\ \hline \end{array} \qquad \begin{array}{r} 783 \\ \times\ 3 \\ \hline \end{array} \qquad \begin{array}{r} 425 \\ \times\ 2 \\ \hline \end{array}$$

B. For more practice, multiply the numbers below. Estimate first.

3. a.

$$\begin{array}{r} 57 \\ \times\ 5 \\ \hline \end{array} \qquad \begin{array}{r} 241 \\ \times\ 9 \\ \hline \end{array} \qquad \begin{array}{r} 308 \\ \times\ 6 \\ \hline \end{array}$$

b.

$$\begin{array}{r} 414 \\ \times\ 4 \\ \hline \end{array} \qquad \begin{array}{r} 39 \\ \times\ 7 \\ \hline \end{array} \qquad \begin{array}{r} 708 \\ \times\ 8 \\ \hline \end{array}$$

4. a.

$$\begin{array}{r} 407 \\ \times\ 6 \\ \hline \end{array} \qquad \begin{array}{r} 3,043 \\ \times\ 3 \\ \hline \end{array} \qquad \begin{array}{r} 6,070 \\ \times\ 8 \\ \hline \end{array}$$

b.

$$\begin{array}{r} 4,009 \\ \times\ 9 \\ \hline \end{array} \qquad \begin{array}{r} 5,081 \\ \times\ 7 \\ \hline \end{array} \qquad \begin{array}{r} 1,006 \\ \times\ 7 \\ \hline \end{array}$$

Regrouping More than Once

Sometimes you have to regroup more than once.

Regrouping More than Once

Example: True Bearings processes about 438 orders a day. In a typical week, how many orders would be processed? $438 \times 5 = ?$

Step 1	Step 2	Step 3	Step 4
Put the larger number on top.	Multiply the ones. Regroup the 4.	Multiply the tens. Add 4. Regroup the 1.	Multiply the hundreds. Add the 1.

Step 1:
```
438
× 5
```

Step 2:
```
  4
438
× 5
___
  0
```

Step 3:
```
1 4
438
× 5
___
 90
```

Step 4:
```
1 4
438
× 5
_____
2,190
```

C. As you do these problems, practice regrouping more than once.

5.
```
 396        4,075       1,453       2,099        634
× 2         ×   8       ×   3       ×   7        × 5
```

6.
```
2,516      10,904        875      24,050        777
×   9      ×     6      × 4       ×     3       × 8
```

The "multiplication computer" below is a different way of approaching multiplication. It receives data through **input.** Then a multiplication occurs and it produces the **output.**

Example:

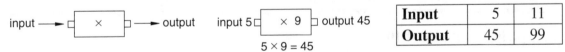

$5 \times 9 = 45$

Input	5	11
Output	45	99

D. Use the multiplication computer to solve problems 7 and 8.

7. From the chart on the right, find the output for each input. Look at the first two inputs and outputs to find the multiplier.

Input	8	10	17	48	175
Output	24	30			

8. From the chart on the right, find the output for each input.

Input	4	9	18	50	376
Output	24	54			

Multiples of 10

Multiplying by 10, 100, or 1,000 is easy. Because 1 times any number is the number itself, just attach the number of zeros found in the multiplier to the number. This will give you the answer.

Example 1: Because $4 \times 1 = 4$:

$4 \times 10 = 40$ ⌐

Attach 1 zero.

$4 \times 100 = 400$ ⌐

Attach 2 zeros.

$4 \times 1,000 = 4,000$ ⌐

Attach 3 zeros.

Example 2: Because $386 \times 1 = 386$:

386	386	386
$\times\ 10$	$\times\ 100$	$\times\ 1,000$
3,860	38,600	386,000

A. Multiply by 10, 100, or 1,000 as indicated.

1. $10 \times 7 =$ $5 \times 10 =$ $100 \times 9 =$ $7 \times 100 =$ $1,000 \times 3 =$

2. $45 \times 10 =$ $10 \times 12 =$ $511 \times 100 =$ $100 \times 481 =$ $94 \times 1,000 =$

Multiples of 10 are numbers that end with zero, such as 20 and 700. When you multiply by multiples of 10, first multiply the nonzero digits. Then attach the total number of zeros.

Examples:

$6 \times 2 = 12$
60 ← 1 zero
$\times\ 20$ ← 1 zero
1,200

⌐ Attach 2 zeros.

$5 \times 3 = 15$
500 ← 2 zeros
$\times\ 30$ ← 1 zero
15,000

⌐ Attach 3 zeros.

$8 \times 9 = 72$
8,000 ← 3 zeros
$\times\ 900$ ← 2 zeros
7,200,000

⌐ Attach 5 zeros.

B. Practice multiplying multiples of 10.

3. a.

50	700	600	**b.** 7,000	80	20
$\times\ 3$	$\times\ 7$	$\times\ 9$	$\times\ \ 8$	$\times\ 6$	$\times\ 5$

4. a.

30	90	400	**b.** 3,000	40	50,000
$\times\ 60$	$\times\ 90$	$\times\ 50$	$\times\ \ 80$	$\times\ 90$	$\times\ \ 70$

5. $231 \times 30 =$ $543 \times 200 =$ $720 \times 40 =$ $1,200 \times 60 =$ $180 \times 50 =$

Estimating Multiplication

Estimating multiplication is a valuable tool. You can use it to find an approximate answer and to check your exact answer.

To estimate multiplication problems
- round each number to a number that's easy to work with
- multiply the nonzero digits
- attach the total number of zeros

Keep in mind that there is no *one* correct estimate. As long as your estimate is reasonable, it is a good estimate.

Rounding to Estimate
Remember: $\approx$ means "is approximately equal to."

Example 1: Estimate 38×11.
Round: $38 \approx 40$ and $11 \approx 10$.

$$\begin{array}{r} 40 \\ \times\ 10 \\ \hline 400 \end{array}$$

40 ← 1 zero
× 10 ← 1 zero
400 ← Attach 2 zeros.

400 is a good estimate of the exact answer **418**.

Example 2: Estimate $6{,}875 \times 521$.
Round: $6{,}875 \approx 7{,}000$ and $521 \approx 500$.

7,000 ← 3 zeros
× 500 ← 2 zeros
3,500,000 ← Attach 5 zeros.

└ 7 × 5 = 35

3,500,000 is a fairly close estimate of the exact answer **3,581,875**.

C. **Estimate an answer to each problem and then choose the exact answer that is closest to your estimate.**

	Estimate		Exact Answer		
6. 89×49	_____		**(1)** 43,061	**(2)** 403,601	**(3)** 4,361
7. 893×62	_____		**(1)** 5,366	**(2)** 55,366	**(3)** 553,366
8. 315×506	_____		**(1)** 15,939	**(2)** 1,539	**(3)** 159,390

D. **Use estimation to give approximate answers to the problems.**

9. Estimate the cost of four $88 tires.

11. Estimate your yearly earnings if you make $625 a week. (There are 52 weeks in a year.)

10. **Explain** If your heart beats about 72 times a minute, estimate the number of beats in an hour. Is your answer higher or lower than the actual answer? Why?

12. **Explain** Estimate the revenue from the sale of 3,899 tickets. Is your estimate higher or lower than the actual answer? Why?

Super Concert
$28

Multiplying by Two-Digit Numbers

When you multiply by a two-digit number, you are really multiplying twice. First multiply by the ones digit to get a partial product. Next multiply by the tens digit to get a second partial product. Then add the two partial products.

You may write multiplication problems in short form by leaving blank spaces for the zeros. Just be careful when you do this.

Multiplying by Larger Numbers

Example 1: 294 × 62   Estimate: 300 × 60 = 18,000

Step 1	**Step 2**	**Short Form**
Put the longer number on top. Multiply to find the partial products.	Add the partial products.	The zero is left off of the second partial product. Begin the partial product in the tens place.

294	294	294
× 62	× 62	× 62
588	588	588
17640	+ 17 640	+ 17 64
	18,228	18,228

Your answer of **18,228** is close to your estimate of 18,000.

Example 2: 48 × 65

Estimate	**Multiply**	**Short Form**
50	48	48
× 70	× 65	× 65
3,500	240	240
	2 880	2 88
	3,120	3,120

Your answer of **3,120** is close to your estimate of 3,500.

A. Practice multiplying by the two-digit numbers below. Estimate first. Then check your answers with your estimate.

1.

743	56	507	152	1,236
× 31	× 25	× 96	× 87	× 44

2. 72 × 198   468 × 53   98 × 1,486   64 × 99

B. For more practice, solve the problems below. Use the short form whenever possible.

3.
$$654 \times 47 \qquad 495 \times 24 \qquad 1,760 \times 26 \qquad 8,901 \times 35 \qquad 64,095 \times 77$$

4. 471 × 33   64 × 908   25 × 25   777 × 18

C. Use your multiplication skills to solve the problems. Estimate first to get a sense of the answer.

5. At the Taste of Chicago, a food vendor sold 18 cases of soda pop. If each case held 24 cans, how many cans of pop did she sell?

6. If you save $25 a week, how much will you save in a year? (One year is 52 weeks.)

7. How many seats are there in a theater that has 36 rows of 48 seats each?

8. Multiplication works both ways. For example, 3 × 9 = 27 and 9 × 3 = 27. Fill in the chart below by performing only one calculation.

×	35	42	18
18	630		324
35	1,225	1,470	
42		1,764	

D. Use the circle graph below to get the information necessary to solve the problems.

9. How much rent does the Shapiro family pay in a year? (One year is 12 months.)

10. If the family follows this budget, can they afford $9,500 in car expenses over 3 years? Why or why not?

11. Multiple Solutions Find the total amount the Shapiros spend in a year following this budget. Is there more than one way to find the answer? If so, how?

Shapiro Family Monthly Budget

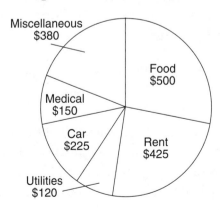

Miscellaneous $380
Medical $150
Car $225
Utilities $120
Food $500
Rent $425

What Information Do I Need?

Problem Solver

When you are solving a problem, read it carefully to select the numbers that will help you answer the question. These numbers are called **necessary information.** Sometimes the numbers are not obvious. Look for words that have values, such as *double* (2), *dozen* (12), *inches in a foot* (12), *triple* (3), *decade* (10), *days in a week* (7).

Can you think of others?

Choosing the Necessary Information

Example 1: If you watch TV about 3 hours daily, how many hours do you watch in a year?

Necessary information: 3 hours and 365 (days in a year)

Some problems contain *extra* information. Be sure to choose only the necessary information.

Example 2: On Monday Maria and her 3 coworkers distributed the new telephone directory to some of the 12,000 households in Parkstown. If each person delivered 265 copies, what was the total distribution that day?

Necessary information: 4 people (Maria + 3 coworkers) and 265 copies each

Extra information: 12,000 households

A. Read the problems and select the necessary information. Write the labels that go with the numbers. *Do not solve.*

1. By three o'clock Monica had spent twice as much money as Erik at the department store. What was Monica's bill if Erik bought only a $38 sweater?

   Necessary information: _____

2. The 16 students in Ahmed's adult education class each wrote a paragraph of 50 words or less describing the number 12. What is the maximum number of words that could be written by the whole class?

   Necessary information: _____

3. If Jack and his 4 children go to Hamburger Haven, Jack estimates that he will spend about $4 per person. Will the $25 in his wallet be enough money to pay the dinner bill?

   Necessary information: _____

4. Rachel took orders for 4 dozen gift boxes this week. The store owners estimated that about 40 boxes would be sold. Did they estimate too many or too few?

   Necessary information: _____

116

B. Solve the problems in Part A.

C. Select the necessary information and solve each problem.

5. Although his boss wants him to make 35 telemarketing calls per hour, Martin averages only 27 calls per hour. At that rate, how many calls does Martin make during his 40-hour workweek?

   Necessary information: _____

6. You are planning your dream vacation. If you stay at the hotel for 2 weeks, what will the hotel bill be?

   Island Paradise Package includes:
   - round-trip airfare: $486
   - food: $70 per day
   - deluxe hotel accommodations: $75/night

   Necessary information: _____

7. Out of 2,080 employees at Alpha Company, 1,682 belong to the union. If each member pays $96 per year, how much money does the union collect?

   Necessary information: _____

8. Anton's delivery trucks each hold 240 cases. If his 8 trucks make 2 deliveries a day, how many cases can 1 truck deliver during a 5-day workweek?

   Necessary information: _____

Making Connections: Buying in Quantity

Sometimes items are cheaper when you buy in larger quantity. Using the price list below, develop a story. Replace the blanks with numbers that make sense. There is no *one* correct story.

BEST BANQUETS

NUMBER OF GUESTS	PRICE PER GUEST
1 TO 9	$45
10 TO 29	$40
30 TO 45	$35
MORE THAN 45	$30

1. Chan and _____ friends attended a charity banquet. His favorite charity is _____. Each banquet ticket cost _____. Chan and his friends spent a total of _____ on the tickets.

Use the price list to answer these questions.

2. a. How much does it cost for 29 guests?
   b. For 30 guests?

3. Explain why you might buy 30 tickets even if only 29 people are going.

Mixed Review

A. Multiply the problems below.

1. a. $\begin{array}{r} 9 \\ \times\ 8 \\ \hline \end{array}$ $\begin{array}{r} 7 \\ \times\ 6 \\ \hline \end{array}$ $\begin{array}{r} 8 \\ \times\ 9 \\ \hline \end{array}$ **b.** $\begin{array}{r} 6 \\ \times\ 6 \\ \hline \end{array}$ $\begin{array}{r} 9 \\ \times\ 7 \\ \hline \end{array}$ $\begin{array}{r} 7 \\ \times\ 8 \\ \hline \end{array}$

2. a. $\begin{array}{r} 9 \\ \times\ 9 \\ \hline \end{array}$ $\begin{array}{r} 6 \\ \times\ 9 \\ \hline \end{array}$ $\begin{array}{r} 8 \\ \times\ 6 \\ \hline \end{array}$ **b.** $\begin{array}{r} 9 \\ \times\ 6 \\ \hline \end{array}$ $\begin{array}{r} 0 \\ \times\ 7 \\ \hline \end{array}$ $\begin{array}{r} 7 \\ \times\ 1 \\ \hline \end{array}$

3. $4 \times 4 =$ $8 \times 7 =$ $3 \times 8 =$ $4 \times 7 =$

4. $8 \times 9 =$ $6 \times 7 =$ $6 \times 4 =$ $5 \times 0 =$

5. $9 \times 10 =$ $3 \times 1{,}000 =$ $6 \times 100 =$ $8 \times 1{,}000 =$

6. a. $\begin{array}{r} 32 \\ \times\ 4 \\ \hline \end{array}$ $\begin{array}{r} 511 \\ \times\ 7 \\ \hline \end{array}$ $\begin{array}{r} 601 \\ \times\ 9 \\ \hline \end{array}$ **b.** $\begin{array}{r} 700 \\ \times\ 8 \\ \hline \end{array}$ $\begin{array}{r} 321 \\ \times\ 3 \\ \hline \end{array}$ $\begin{array}{r} 902 \\ \times\ 4 \\ \hline \end{array}$

7. $3 \times 8 \times 2 =$ $6 \times 5 \times 7 =$ $9 \times 4 \times 0 =$

B. Multiply to solve these problems. Estimate first, then compare your answer to your estimate.

8. a. $\begin{array}{r} 85 \\ \times\ 9 \\ \hline \end{array}$ $\begin{array}{r} 37 \\ \times\ 6 \\ \hline \end{array}$ $\begin{array}{r} 79 \\ \times\ 4 \\ \hline \end{array}$ **b.** $\begin{array}{r} 1{,}278 \\ \times\ 4 \\ \hline \end{array}$ $\begin{array}{r} 6{,}007 \\ \times\ 8 \\ \hline \end{array}$ $\begin{array}{r} 909 \\ \times\ 9 \\ \hline \end{array}$

9. $\begin{array}{r} 40 \\ \times\ 80 \\ \hline \end{array}$ $\begin{array}{r} 300 \\ \times\ 90 \\ \hline \end{array}$ $\begin{array}{r} 485 \\ \times\ 20 \\ \hline \end{array}$ $\begin{array}{r} 607 \\ \times\ 500 \\ \hline \end{array}$ $\begin{array}{r} 92{,}000 \\ \times\ 30 \\ \hline \end{array}$

10. $\begin{array}{r} 26 \\ \times\ 55 \\ \hline \end{array}$ $\begin{array}{r} 906 \\ \times\ 57 \\ \hline \end{array}$ $\begin{array}{r} 480 \\ \times\ 38 \\ \hline \end{array}$ $\begin{array}{r} 564 \\ \times\ 32 \\ \hline \end{array}$ $\begin{array}{r} 4{,}806 \\ \times\ 77 \\ \hline \end{array}$

11. $847 \times 20 =$ $254 \times 37 =$ $1{,}075 \times 16 =$

12. $306 \times 9 =$ $900 \times 40 =$ $346 \times 81 =$

C. Use your multiplication skills to solve the following problems.

13. You drive about 175 miles each week for a year. What is your total annual mileage?

14. Darryl makes $15 per hour and double time for every hour over 40 hours per week. If he worked 45 hours last week, what was his total pay?

15. Find the value of the variable in each equation below.

$5k = 45$ $8(7) = H$ $6t = 54$

$k =$ $H =$ $t =$

Hint: $5 \times$ ___ $= 45$

16. Ninety-eight crafters paid $37 each to display their wares at the holiday craft show. If the money was donated to the local homeless shelter, what was the total donation?

D. Discuss this bar graph with a partner. Then use the information to answer the following questions.

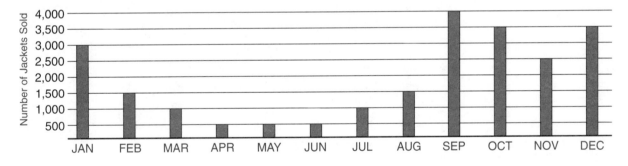

17. The Just Jackets Company has set a goal to triple its August sales next year. To do this, how many jackets will they have to sell?

18. If the jackets sell for $68 each, how much money did the company receive in September from the sale of jackets?

19. a. At a sale in January the jacket sold for $51, but in November the jacket sold for the full price of $68. In which month did the company make more money?

b. How much money did it make in sales that month?

20. Investigate Is there any extra information in the graph that you have not used? Explain your answer.

Multiplying by Larger Numbers

When you multiply by larger numbers, you get a partial product for each digit in the multiplier.

Working with Partial Products

Example 1: 321×264

Estimate first.	Multiply to find the partial products. Then add.	Or use the short form.
300 × 300 90,000	321 × 264 1 284 19 260 64 200 84,744	321 × 264 1 284 19 26 64 2 84,744

Your answer of **84,744** is close to your estimate of 90,000.

If a partial product is all zeros, you only need to write a zero to hold the same place value as the digit in the multiplier. Then write the next partial product on the same line in the short form.

Example 2: 567×304

Estimate first.	Multiply to find the partial products. Then add.	Short form.
600 × 300 180,000	567 × 304 2 268 0 000 170 100 172,368	567 × 304 2 268 170 10 172,368

Your answer of **172,368** is close to your estimate of 180,000.

A. Estimate first. Then find an exact answer.

1. 356×183 Estimate *350* *× 200* 152×875 $1,486 \times 724$ $8,901 \times 365$

2. 654×407 750×909 $3,045 \times 503$ $18,301 \times 1,005$

3. The nearest professional baseball stadium seats 42,365 people. If it is filled for each game, how many spectators attended the 162 games played during the last two seasons?

B. Practice your multiplication skills with this cross number puzzle.
(*Hint:* First fill in the answers you can do quickly; see 23 down.)

Across

1. 32 × 11

4. 24 × 26

7. 8 × 7

9. 63 × 10

10. 61 × 7

11. 1 × 13

12. 15 × 19

14. 75 × 94

16. 13 × 5

17. 10 × 20

18. 9 × 9

20. 5 × 10

22. 25 × 17

25. 9 × 83

27. 4 × 4

29. 22 × 2

30. 57 × 9

33. 5 × 9

35. 203 × 30

36. 91 × 9

38. 10 × 4

39. 15 × 25

41. 3 × 68

43. 2 × 45

44. 7 × 92

45. 127 × 4

Down

1. 4 × 9

2. 28 × 19

3. 2 × 1,043

4. 8 × 8

5. 8 × 284

6. 47 × 100

7. 17 × 30

8. 9 × 7

13. 5 × 111

15. 9 × 56

18. 29 × 3

19. 7 × 2

21. 8 × 0

23. 6 × 4

24. 9 × 6

26. 15 × 50

27. 1 × 1

28. 12 × 54

31. 44 × 44

32. 58 × 53

34. 1,025 × 5

35. 6 × 100

37. 30 × 30

38. 7 × 7

40. 6 × 9

42. 8 × 6

Area and Volume

Finding Area

The **area** of a square or a rectangle is measured by the number of **square units** that fit on the surface.

This is one square centimeter.

This is one square inch.

Finding Area

Example: Suppose you have a rectangle that is 3 centimeters wide and 4 centimeters long.

You can see that there are 12 square centimeters that cover the surface.

To find the surface area, multiply the length by the width. In this example, 3 centimeters × 4 centimeters = **12 square centimeters.**

3 centimeters

4 centimeters

Area measures can be used to find out the amount of surface a floor or a wall has. If you want to tile a kitchen floor, for instance, first find the size of its surface area. For a floor that is 15 feet on each side, multiply the length by the width.

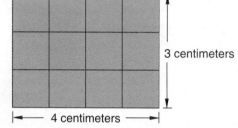

15 feet

15 feet

15 feet × 15 feet = 225 square feet

It is much easier to multiply the length by the width than to count the squares in the picture.

▶ The formula for finding the area of a square or rectangle is:
Area = length × width
$A = l \times w$

A. Use your knowledge of area to solve the problems below.

1. Explain why all three figures on the right have an area of 12 square units.

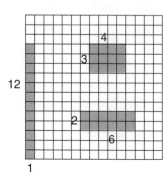

2. Draw three figures that each have an area of 16 square units.

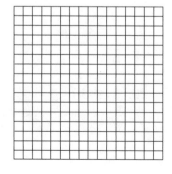

3. Find the area of each figure below.

a.
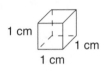
7 in.

7 in.

b.
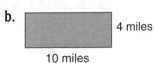
4 miles

10 miles

c.
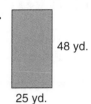
48 yd.

25 yd.

4. How much sod do you need to buy to cover a football field 120 yards long and 55 yards wide?

Finding Volume

Volume is the amount of space inside a figure that has length, width, and height. It is always measured in cubic units.

1 cubic centimeter

1 cm

1 cm

1 cm

A **cubic unit** is a three-dimensional unit used to measure the volume of figures such as cubes or rectangular boxes.

This is a rectangular box. It contains **30 cubic centimeters.**

We can find the volume by multiplying the length by the width by the height.

5 cm × 2 cm × 3 cm = **30 cubic centimeters**

3 cm

2 cm

5 cm

▶ The formula for finding the volume of a cube or rectangular box is:
   Volume = length × width × height $V = l \times w \times h$

For another look
at formulas, turn to
page 205.

B. Use your knowledge of volume to solve the problems below.

5. Find the volume of the figure.

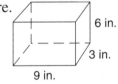
6 in.
3 in.
9 in.

7. Will a furnace designed to heat 7,500 cubic feet be enough to heat an apartment with 8-foot-high ceilings that is 30 feet long and 25 feet wide?

6. a. Find the volume of a swimming pool with these measurements.

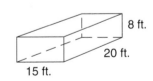
8 ft.
20 ft.
15 ft.

8. **Write** Mark length, width, and height on the box below. Then write and solve your own problem.

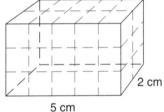

b. If there are about 7 gallons of water in one cubic foot, how many gallons of water will it take to fill up the pool?

Multiplying Dollars and Cents

You often have to multiply dollars and cents in daily life. Here are some tips to follow when you multiply money.

Tip 1. When you multiply dollars, line up the digits on the right. You do not need a decimal point if there are no cents.

Example: $63 × 4

$63
× 4
$252 ◀—— This means the same as $252.00

Tip 2. When you multiply dollars and cents, line up the digits on the right. Multiply the numbers. Put the decimal point 2 places to the left in the final answer.

Example: $52.34 × 25

$52.34
× 25
261 70 ◀—— Do not use decimal points in partial
1046 8 ◀—— products. Place a $ and a decimal
$1,308.50 ⬋ point in the answer.

Tip 3. When you multiply cents and the answer is more than 99¢, you must use the $ and decimal point.

Example: 75¢ × 5 —▶ 75¢ = $.75

$.75
× 5
$3.75

A. Multiply the amounts below.

1. $36 × 7 $500 × 20 $2,725 × 16

2. $7.95 × 6 $680.50 × 60   48 × $219.46

3. 35¢ × 2   89¢ × 24   60¢ × 40

B. For more practice, multiply the following money amounts.

4. $98 × 4 $742 × 15 $13,065 × 30

5. $15.45 × 3 $67.08 × 32   77 × $205.60

6. 95¢ × 36   9¢ × 100   25¢ × 72

C. Use your multiplication skills to solve the following.

7. In a special promotion, a grocery store gave a triple coupon discount. If Fumiko has a 75¢ coupon, what is her discount?

8. Jawon pays $278.35 per month to lease a car. How much will he pay if his lease lasts 2 years?

9. Paula and Ana Maria sold 50 flower arrangements at the craft show. If each one sold for $24.99, how much money did they collect?

10. John delivers daily newspapers to 64 households. At the end of each month he collects $18.50 from each newspaper customer. Find the total amount he collects.

Making Connections: Taking Inventory

Every January, Eagle Sport Shop takes an inventory of all the sporting goods in the store. Employees list the items, tally the number of items, and multiply the total by each item's value. Sample items are listed below. Complete the chart. The first row is done for you.

Eagle Sport Shop Inventory: January

Item	Tally	Total Items	Value per Item	Total Amount
Headgear	|||| ||||| ||||	14	$11.45	$160.30
Sport socks	|||| |||| |||| |||| |||		3.68	
Jump ropes	|||| |||| |||| |||		1.95	
Sport gum	|||| |||| |||| |||| |||| ||||		.37	
Sweat suits	|||| |||| |||| |||| ||		29.15	
Treadmill	|||| |||| ||||		138.60	

Total Value: _____

What is the total value of this inventory?

Multiplying on a Calculator

The steps to perform the multiplication operation on a calculator are

- press On/Clear
- enter the first number
- press the multiplication key $\boxed{\times}$
- enter the next number
- press the equal sign $\boxed{=}$

Remember to estimate your answer to double-check your calculator answer. Always ask yourself, "Does my answer make sense?"

Using a Calculator

Example: 573 × 942

Key in:	Your display reads:	Estimate
$\boxed{\frac{On}{C}}$	0.	600
$\boxed{5}\boxed{7}\boxed{3}$	573.	$\times\ 900$
$\boxed{\times}$	573.	540,000
$\boxed{9}\boxed{4}\boxed{2}$	942.	
$\boxed{=}$	539766.	

Practice this procedure on your calculator.

Here are a few tips to keep in mind when you use a calculator for multiplication.

Tip 1. When you multiply three or more numbers in a row, enter each number followed by the multiplication key until you enter the last number. Then press the equal sign.

Example: 48 × 19 × 346

Key in:	Your display reads:	Estimate
$\boxed{\frac{On}{C}}$	0.	50
$\boxed{4}\boxed{8}$	48.	$\times\ 20$
$\boxed{\times}$	48.	1,000
$\boxed{1}\boxed{9}$	19.	$\times\ 300$
$\boxed{\times}$	912.	300,000
$\boxed{3}\boxed{4}\boxed{6}$	346.	
$\boxed{=}$	315552.	

A. Use a calculator to solve the problems below. Estimate first.

1. 396 × 475	2,914 × 309	17,462 × 218
2. 78 × 566	6,032 × 64	724 × 907
3. 27 × 65 × 39	486 × 12 × 42	8 × 84 × 31
4. $68.45 × 18	98¢ × 125	$479.28 × 48

B. For more calculator practice, solve these problems. Estimate first.

5. 927 × 78

Estimate: _____

Actual: _____

7,165 × 344

Estimate: _____

Actual: _____

6. 63 × 24 × 27

Estimate: _____

Actual: _____

155 × 9 × 48

Estimate: _____

Actual: _____

C. Jeremy has decided to calculate his yearly expenses. Use your calculator to help him find the totals for the expenses below.

Item	Amount	Yearly Total
Rent	$386 per month	_____
Food	$75 per week	_____
Car insurance	$1,064 per year	_____
Health insurance	$28 per week	_____
Utilities	$150 per month	_____
Car payments	$247 per month	_____
Gas	$96 per month	_____
Phone	$35 per month	_____
Miscellaneous	$100 per week	_____

Total: _____

Multistep Problems

Although we have not discussed multistep problem solving, you have solved some
problems that required you to use more than one operation. Let's look at a couple
of problems that require you to use a combination of addition, subtraction, and
multiplication.

Solving Multistep Problems

Example 1: Mr. Moore makes $11.45 per hour and Mrs. Esposito makes $13.50
per hour. How much more does Mrs. Esposito earn than Mr. Moore earns during a
40-hour week?

Step 1
Multiply to find each person's earnings.

Mr. Moore	Mrs. Esposito
$11.45	$13.50
× 40	× 40
$458.00	$540.00

Step 2
Subtract to find how much more Mrs. Esposito earns.

$540.00
− 458.00
$82.00

Example 2: Find the total surface area in the floor plan on the right.

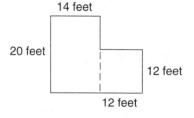

Step 1
Separate the square from the rectangle. Find
the area of each figure (length × width).

Rectangle	Square
14	12
× 20	× 12
280 square feet (sq. ft.)	144 square feet (sq. ft.)

Step 2
Add to find the
total area.
  280 sq. ft.
+ 144 sq. ft.
  424 sq. ft.

A. Practice solving the multistep problems below.

1. If Gloria earns $540 per week and Ramon earns $12.45 per hour for a 40-hour
   week, what are their total wages for the week?

   Step 1:

   Step 2:

2. If Ramon gets a raise of $1.86 per hour, will he earn more or less than Gloria in a
   40-hour work week? Explain your reasoning.

B. Use your geometry skills to solve these problems.

> Remember that Area = length × width

3. Find the total surface area of the floor plan shown.

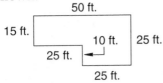

4. If a flooring contractor charges $3 per square foot to refinish a wood floor, what is the cost to refinish a room measuring 18 feet by 24 feet?

5. **Explain** If 1 gallon of sealant covers 300 square feet, will 20 gallons of sealant be enough to cover a square warehouse floor that is 75 feet on each side? Explain your answer. Draw a picture if you need to.

C. Here is a birthday exercise that you can show to your friends. You may use your calculator.

1. Enter the number of the month you were born.   _____

2. Multiply by 20.   × 20

3. Add 100 to the answer.   + 100

4. Multiply this answer by 5.   × 5

5. Now subtract 375.   − 375

6. Add the day of the month you were born.   + ____

7. Multiply by 4.   × 4

8. Subtract 80.   − 80

9. Now multiply by 25.   × 25

10. Add the last two digits of your birth year.   + ____

11. Subtract a whole number greater than 10,499 and less than 10,501.   − ____

This is your birthday: _____
                                month/day/year

Now this is a multistep problem!

Filling Out an Order Form

Many people enjoy buying things through catalogs. Filling out an order form organizes the items you want to buy and lets you figure how much they will cost.

The catalog number, item description, and price per item have been filled in on the sample order form below. Multiply the quantity ordered by the price for each item to get the price total. Finally, use addition to complete the form.

A. Complete the order form. You may use a calculator.

Country Home Shop

Catalog Number	Item Description	Quantity	Price Each	Price Total
A770-3366	Table lamps	2	49.99	_____
A745-4051	Woven rug	1	24.60	_____
A736-8707	Mini-blinds	4	18.40	_____
A727-3311	Pillow	2	19.99	_____

Merchandise Total: _____

Sales Tax: 16.67

Delivery/Shipping Charges: 3.95

Total: _____

B. An order form for clothes usually requires more information, such as color and size of the item. Use the order form below to answer the questions on page 131. You may use a calculator.

Catalina Casuals

ITEM NUMBER	ITEM DESCRIPTION	COLOR	SIZE	QUANTITY	PRICE EACH	PRICE TOTAL
PP15804	Belt	Brown	36	1	$12.00	$12.00
PP27215	Cotton tees	White, black, blue	XL	3	7.99	
PP27409	Denim shorts	Blue	34	2	19.99	
PP27214	Jacket	Khaki	XL	1	39.99	39.99
PP01095	Moccasins	Brown	11M	1	32.00	32.00

Merchandise: _____

Tax: 10.36

Shipping: 7.19

Total: _____

1. Before tax, how much will it cost to buy 3 cotton tees and 2 pairs of denim shorts?

2. Tax and shipping are extra charges. How much are they altogether?

3. What is the total cost for this order?

4. If you decide not to order the jacket, what is the *new merchandise total?*

C. Fill out the forms below. You may use a calculator.

5. If you have only $75 to spend, which items from the order form on page 130 would you select to purchase? Get as close to $75 as possible. Include a tax of $5.60 and $4.75 for shipping.

ITEM NUMBER	ITEM DESCRIPTION	COLOR	SIZE	QUANTITY	PRICE EACH	PRICE TOTAL

Merchandise:_____

Tax:_____

Shipping:_____

Total:_____

6. Fill out the order form below using the prices at right. Order the following tickets:
   July 17: 4 adults, 1 student, 3 children
   August 3: 10 seniors, 2 adults

Child (under 12)	$12.50
Student	$15.00
Adult	$20.00
Senior (over 60)	$14.00

Popular Productions Order Form

Date	Ticket Description	Quantity	Price Each	Price Total

Ticket Total:_____

Shipping/Handling:____3.50____

Total:_____

Unit 4 Review

A. What number will make each statement true?

1. _____ is the largest number you can make from the digits 5, 3, 7, and 9, using each one once.

2. _____ is an even number between 386 and 390.

3. _____ feet equal 1 yard.

4. $148.25 rounded to the nearest $10 is _____.

5. $48 + \underline{\hspace{1cm}} = 48$

6. $\underline{\hspace{1cm}} > 95$

7. Twenty thousand forty-six can also be written as _____.

8. $38¢$ plus $95¢ = \underline{\hspace{1cm}}$

9. There are _____ hours in a day.

10. $\$297 + \$34.25 = \underline{\hspace{1cm}}$

11. The next number in the pattern 4, 7, 10 is _____.

12. _____ is the perimeter of a square that is 7 inches on each side.

13. If $97 + 36 = 133$, then $133 - 97 = \underline{\hspace{1cm}}$.

14. Seventeen dollars and six cents can also be written as _____.

15. $17,000 + 3,050 + 194 = \underline{\hspace{1cm}}$

16. $456 - 157 = \underline{\hspace{1cm}}$

17. $2,000 - 723 = \underline{\hspace{1cm}}$

18. $\$60.10 - \$3.89 = \underline{\hspace{1cm}}$

19. If $75 - C = 35$, then $C = \underline{\hspace{1cm}}$.

20. If $y + y = 12$, then $y = \underline{\hspace{1cm}}$.

B. Use your multiplication and estimation skills below.

21. a.
$$\begin{array}{cccc} 7 & 8 & 9 & 8 \\ \times\,6 & \times\,7 & \times\,6 & \times\,9 \end{array}$$
b.
$$\begin{array}{cccc} 7 & 7 & 6 & 7 \\ \times\,7 & \times\,9 & \times\,8 & \times\,8 \end{array}$$

22. a.
$$\begin{array}{ccc} 44 & 127 & 55 \\ \times\,2 & \times\,4 & \times\,6 \end{array}$$
b.
$$\begin{array}{ccc} 8,004 & 6,724 & 478 \\ \times\,6 & \times\,8 & \times\,600 \end{array}$$

23. $50 \times 30 = \qquad 2,275 \times 80 = \qquad 2,046 \times 9,000 = \qquad 2,008 \times 720 =$

24. $65 \times 45 = \qquad 3,163 \times 27 = \qquad 4,573 \times 604 = \qquad 2,030 \times 356 =$

C. Solve the problems.

25. If Express Air delivers 1,036 packages per week, how many are delivered in a year?

26. How much change will you get back from two $20 bills if you buy a dozen roses at $2.35 each?

27. Nana cut the lasagna shown below into 24 pieces. Show two other ways of cutting 24 equal pieces.

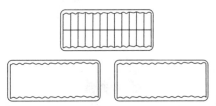

28. How many square feet of floor tile do you need to cover the family room floor shown below?

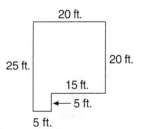

20 ft.
25 ft.
20 ft.
15 ft.
5 ft.
5 ft.

Working Together

Popular Productions organizes the 12-week summer concert series. They sell tickets for the prices shown below. Prices include tax.

Ticket	Tuesday/Thursday Night	Wednesday Matinee	Friday/Saturday Night	Sunday Matinee
Adult	$26.00	$20.00	$32.00	$30.00
Senior (over 60)	18.20	14.00	22.40	21.00
Student (12–18)	19.50	15.00	24.00	22.50
Child (under 12)	14.75	12.50	20.00	18.75

With a partner, fill out the order form below for three summer dates. You may want to take family members and friends to the concert, so be sure to order their tickets too.

Popular Productions Order Form

Date	Ticket Description	Quantity	Price Each	Price Total

Ticket Total: _____

Shipping/Handling: 3.50

Total: _____

Unit 5

Division

Skills

Counting

Dividing numbers

Working with remainders

Dividing by multiples of 10

Dividing money

Tools

Number line

Multiplication table

Calculator

Problem Solvers

Division strategies

Item sets

Choosing the right operation

Applications

Finding an average

Finding unit price

Suppose 3 friends decide to equally share the expenses of an apartment. If the monthly rent is $630, what is each person's rent payment?

Now suppose you borrow an amount of money and plan to repay the loan in monthly installments. How long will it take to pay off $630 if you make $90 loan payments?

How do you find the answers to these problems? You divide.

Division is the operation you use to

- separate an amount into equal parts
- find how many times one amount fits into another amount

134

When Do I Divide?

Words or phrases we can use to indicate division are

share equally	how many times
split	go into
each	divide into
average	how many measured amounts fit in another

Think about different times in your life when you might use division to

1. split an amount into equal parts

2. find how many times one amount fits into a larger amount

There are three symbols that indicate division. The problem "eighteen divided by three equals six" can be written in these three ways.

division bracket

$$3\overline{)18}^{6}$$

division sign

$$18 \div 3 = 6$$

fraction bar

$$\frac{18}{3} = 6$$

Notice the placement of the 3. It is called the **divisor** (the number you *divide by*). The answer, 6, is called the **quotient.** The number being divided is the **dividend.** (In this example, the dividend is 18.)

Identify the dividend, divisor, and quotient, as indicated, in each of the following problems.

3. $24 \div 6 = 4$ divisor: _____ quotient: _____

4. $7\overline{)56}^{8}$ divisor: _____ quotient: _____

5. $\frac{72}{8} = 9$ divisor: _____ quotient: _____

6. $9\overline{)54}^{6}$ dividend: _____ divisor: _____

7. $\frac{28}{4} = 7$ dividend: _____ divisor: _____

8. $40 \div 8 = 5$ dividend: _____ divisor: _____

Talk About It

Discuss the division problems below with a partner. Do not solve the problems. Write each division problem in three different ways. Think of a real-life situation for each problem.

- Split 36 into 9 equal parts.

- How many 6-inch pieces can be cut from a 42-inch board?

- Find 45 divided by 9.

- How many times does 3 go into 12?

Division Strategies

In this chapter you will use your skills with multiplication, subtraction, and educated guessing to find the answer to division problems.

Division is the opposite operation of multiplication. You can use the multiplication table in reverse to help you learn **division facts.**

×	1	2	3	4
1	1	2	3	4
2	2	4	6	8
3	3	6	9	12
4	4	8	12	16

Given the fact that $3 \times 4 = 12$, the related division facts are $12 \div 4 = 3$ and $12 \div 3 = 4$.

A. Write two related division facts for each multiplication fact. The first problem is started for you.

1. $8 \times 3 = 24$ $9 \times 5 = 45$ $7 \times 8 = 56$
   $24 \div 8 =$
   $24 \div 3 =$

2. $6 \times 8 = 48$ $8 \times 4 = 32$ $7 \times 9 = 63$

You can visualize division on the number line. Below is the picture of $12 \div 4 = 3$. You see that 4 can be subtracted from 12 three times.

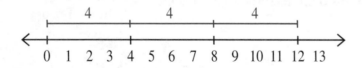

B. Write the division problem pictured on each number line below.

3.

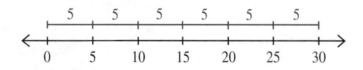

4.

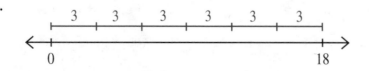

For another look at the multiplication table, turn to page 204.

Division Tips

Memorize the division facts. The multiplication facts and number line will help you learn them. The following tips also will help you.

Tip 1. A number divided by 1 is the number.

    Examples: $8 \div 1 = 8$ $1\overline{)9}$ with quotient 9 $\frac{6}{1} = 6$

Tip 2. Any number divided by itself is 1.

    Examples: $7 \div 7 = 1$ $3\overline{)3}$ with quotient 1 $\frac{4}{4} = 1$

Tip 3. Zero divided by any number is zero.

    Examples: $0 \div 4 = 0$ $5\overline{)0}$ with quotient 0 $\frac{0}{9} = 0$

Tip 4. Check an answer to a division problem by multiplying the quotient by the divisor. Your result should be the dividend.

    Examples: $15 \div 3 = 5$ because $3 \times 5 = 15$

        $9\overline{)81}$ with quotient 9 because $9 \times 9 = 81$

        $\frac{42}{7} = 6$ because $6 \times 7 = 42$

C. Use the division tips above to solve these problems.

5. $9 \div 9 =$ $8 \div 1 =$ $4 \div 4 =$ $0 \div 3 =$

6. $\frac{7}{1} =$ $\frac{0}{2} =$ $\frac{7}{7} =$ $\frac{6}{6} =$

7. $2\overline{)2}$ $1\overline{)3}$ $1\overline{)4}$ $8\overline{)8}$

8. _____ $\div 3 = 0$ _____ $\div$ _____ $= 1$ $8 \div$ _____ $= 8$

D. Use your knowledge of the multiplication facts or the multiplication table on page 204 to help you complete these division facts. If a problem has two blanks, use the same number for both blanks.

9. $16 \div$ _____ $= 8$ _____ $\div 7 = 8$ $64 \div$ _____ $=$ _____

10. _____ $\div 3 = 9$ _____ $\div 5 = 5$ $48 \div$ _____ $= 6$

11. _____ $\div 9 = 0$ $7 \div$ _____ $= 1$ $49 \div$ _____ $=$ _____

Division Facts

As you have seen, you can learn your division facts based on the multiplication facts you already know. It is important to be able to do the division as well as the multiplication facts quickly and accurately.

A. Practice your one-digit division facts on the problems below. Practice these problems until you can do them quickly with no errors.

1. **a.** $18 \div 6 =$ $8 \div 2 =$ $48 \div 6 =$ **b.** $10 \div 5 =$ $56 \div 8 =$ $12 \div 3 =$

2. **a.** $49 \div 7 =$ $0 \div 4 =$ $64 \div 8 =$ **b.** $14 \div 2 =$ $0 \div 3 =$ $6 \div 6 =$

3. **a.** $20 \div 5 =$ $6 \div 3 =$ $30 \div 5 =$ **b.** $28 \div 4 =$ $63 \div 9 =$ $20 \div 4 =$

4. **a.** $21 \div 7 =$ $14 \div 7 =$ $42 \div 6 =$ **b.** $2 \div 1 =$ $81 \div 9 =$ $72 \div 9 =$

5. **a.** $8 \div 1 =$ $1 \div 1 =$ $24 \div 8 =$ **b.** $0 \div 9 =$ $56 \div 7 =$ $3 \div 3 =$

6. **a.** $36 \div 6 =$ $35 \div 7 =$ $48 \div 8 =$ **b.** $15 \div 5 =$ $0 \div 7 =$ $0 \div 6 =$

7. **a.** $40 \div 5 =$ $24 \div 4 =$ $27 \div 9 =$ **b.** $4 \div 4 =$ $54 \div 9 =$ $8 \div 8 =$

8. **a.** $81 \div 9 =$ $9 \div 3 =$ $54 \div 6 =$ **b.** $21 \div 3 =$ $24 \div 6 =$ $36 \div 9 =$

9. **a.** $15 \div 3 =$ $27 \div 3 =$ $25 \div 5 =$ **b.** $12 \div 2 =$ $42 \div 7 =$ $5 \div 1 =$

10. **a.** $35 \div 5 =$ $6 \div 2 =$ $12 \div 4 =$ **b.** $4 \div 1 =$ $36 \div 4 =$ $18 \div 9 =$

11. **a.** $28 \div 7 =$ $0 \div 5 =$ $45 \div 5 =$ **b.** $16 \div 2 =$ $63 \div 7 =$ $12 \div 6 =$

12. **a.** $72 \div 8 =$ $45 \div 9 =$ $16 \div 4 =$ **b.** $7 \div 7 =$ $50 \div 5 =$ $6 \div 1 =$

13. **a.** $32 \div 8 =$ $3 \div 1 =$ $18 \div 2 =$ **b.** $7 \div 1 =$ $24 \div 3 =$ $0 \div 2 =$

14. **a.** $21 \div 3 =$ $9 \div 9 =$ $9 \div 1 =$ **b.** $4 \div 2 =$ $18 \div 3 =$ $30 \div 6 =$

15. **a.** $16 \div 8 =$ $8 \div 4 =$ $32 \div 4 =$ **b.** $0 \div 1 =$ $40 \div 8 =$ $10 \div 2 =$

Division Equations

A **division equation** is a number sentence that represents division. The most common way to write a division equation is to use a fraction bar and equal sign.

Example: "Sixteen divided by two equals eight" is written as

$\frac{16}{2} = 8$ This means the same as 16 ÷ 2 = 8.

Be sure the divisor is written *below* the fraction bar.

B. Write the following sentences as division equations.

16. Twenty split evenly by four is five.

17. Seven goes into twenty-one three times.

18. Six is the quotient of eighteen divided by three.

19. Fifty-four divided by six is nine.

If the division equation contains a variable, remember that the divisor is below the fraction bar.

C. Find the value of the variable in each sentence below.

20. $\frac{36}{9} = m$ $\frac{28}{t} = 4$ $\frac{C}{7} = 8$

   $m \ =$ $t \ =$ $C \ =$

                  Hint: 28 ÷ _____ = 4

21. $\frac{27}{y} = 3$ $\frac{X}{7} = 7$ $\frac{64}{8} = r$

   $y \ =$ $X \ =$ $r \ =$

D. Use your knowledge of the four operations to solve the equations below.

22. $B + 6 = 15$ $4n = 32$ $\frac{70}{W} = 10$

   $B \ =$ $n \ =$ $W \ =$

                  Hint: 4 × _____ = 32

23. $29 - A = 21$ $\frac{E}{8} = 9$ $7f = 56$

   $A \ =$ $E \ =$ $f \ =$

Dividing by One Digit

Long division is a four-step process that repeats until there are no digits left to bring down. Every time you bring down a digit, divide and write in the quotient.

> ▶ **Four-Step Division Process**
>
> **Step 1.** Divide and write the answer above.
>
> **Step 2.** Multiply the answer times the divisor.
>
> **Step 3.** *Subtract.*
>
> **Step 4.** Bring down the next digit and *repeat* the process if needed.

Using the Four-Step Process

Example: 78 ÷ 3   Estimate: 90 ÷ 3 = 30

Step 1
Divide.

$$\begin{array}{r} 2 \\ 3\overline{)78} \end{array}$$

Begin by dividing the left digit. How many times does 3 go into 7? Write the answer above the 7.

Step 2
Multiply.

$$\begin{array}{r} 2 \\ 3\overline{)78} \\ 6 \end{array}$$

2 × 3 = 6
Align the 6 under the 7.

Step 3
Subtract.

$$\begin{array}{r} 2 \\ 3\overline{)78} \\ -6 \\ \hline 1 \end{array}$$

Step 4
Bring down the next digit. Repeat the process.

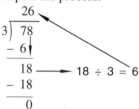

18 ÷ 3 = 6

Your answer of **26** is close to your estimate of 30.

A. Follow the steps shown above as you practice dividing. Estimate first.

1. 48 ÷ 4   Estimate   95 ÷ 5   847 ÷ 7   7,342 ÷ 2

2. 655 ÷ 5   84 ÷ 3   712 ÷ 4   9,512 ÷ 8

Repeating the Four-Step Process

Example: 2,418 ÷ 3   Repeat the four-step process as needed.

Divide

$$\begin{array}{r} 8 \\ 3\overline{)2,418} \\ -24 \\ \hline 0 \end{array}$$

How many 3s in 24?
Put 8 above the 4.
Multiply: 3 × 8 = 24.
Subtract: 24 − 24 = 0.

Repeat

$$\begin{array}{r} 80 \\ 3\overline{)2,418} \\ -24 \\ \hline 01 \\ -0 \\ \hline 1 \end{array}$$

Bring down the 1.
How many 3s in 1?
Put 0 above the 1.
Multiply: 0 × 3 = 0.
Subtract: 1 − 0 = 1.

Repeat

$$\begin{array}{r} 806 \\ 3\overline{)2,418} \\ -24 \\ \hline 01 \\ -0 \\ \hline 18 \\ -18 \\ \hline 0 \end{array}$$

Bring down the 8.
How many 3s in 18?
Put 6 above the 8.
Multiply: 6 × 3 = 18.
Subtract: 18 − 18 = 0.

B. Practice your division skills on the problems below.

3. $4\overline{)348}$ $9\overline{)702}$ $3\overline{)195}$ $7\overline{)182}$ $6\overline{)360}$

4. $2\overline{)2,064}$ $8\overline{)6,432}$ $5\overline{)6,080}$ $9\overline{)2,718}$ $6\overline{)2,946}$

C. Solve the problems below. Estimate first to get a sense of the answer.

5. Marlene ordered 132 balloons for the banquet decorations. If each centerpiece uses 6 balloons, how many tables can she decorate?

6. Lucy and 3 coworkers bought a lottery ticket together. They were disappointed they did not win the $2,560 prize. If they had won, how much would each person have gotten?

7. A total of 162 people signed up to play softball at the company picnic. If there are 9 players on each team, how many teams can be formed?

8. According to the picture below, the Statue of Liberty (including base) is how many times as tall as Jay?

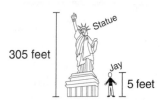

Short Division

Short division is a shortcut that comes in handy when you divide by a one-digit number. Save time by performing the multiplication and subtraction steps mentally.

Examples: $2,807 \div 7$

$$
\begin{array}{r} 401 \\ 7\overline{)2,807} \end{array}
$$

$28 \div 7 = 4$
$0 \div 7 = 0$
$7 \div 7 = 1$

$3,815 \div 5$

$$
\begin{array}{r} 763 \\ 5\overline{)3,815} \end{array}
$$

$38 \div 5 = 7$ (with 3 left over)
$31 \div 5 = 6$ (with 1 left over)
$15 \div 5 = 3$

D. Use short division to find the answers to these problems.

9. $6\overline{)606}$ $2\overline{)9,814}$ $4\overline{)1,616}$ $9\overline{)288}$

10. $8\overline{)856}$ $3\overline{)7,206}$ $5\overline{)19,155}$ $7\overline{)49,105}$

Remainders

A **remainder** in division is the nonzero amount left over from the final subtraction.

Finding the Remainder

Example: An art dealer wants to separate a group of 74 paintings into 3 different displays. Will he be able to display the paintings in 3 equal groups?

```
    24 R2  ◄──────  The letter R stands for remainder. Write the letter before
3) 74                the remaining amount.
 – 6
   14
 – 12
    2
```

The art dealer can display 3 groups of 24 paintings. If the art dealer wants an equal number of paintings in each display, he will need to set aside the 2 remaining paintings.

The remainder *must always be less than* the divisor. If it is not, recheck and correct your division.

To check a division problem multiply the answer times the divisor. Then add the remainder. The final answer should be the same as the number divided.

Checking Your Work

Example: 368 ÷ 6   Estimate: 360 ÷ 6 = 60

```
    61 R2 ◄            Check
6) 368
 – 36                  61              366
    8                 × 6             + 2  ◄─── remainder
  – 6                 366             368
    2 ─┘
```

A. Estimate first and then find an exact answer. Watch for remainders. Check your work.

1. 4)33   6)28   9)421   6)565   7)849

2. 7,990 ÷ 3   3,523 ÷ 2   2,134 ÷ 5   42,951 ÷ 6

Using the Remainder

Example: Sometimes you need to decide what to do with the remainder. For example, the senior center uses vans for sight-seeing trips. If each van seats 9 passengers, how many vans will be needed to take 24 people to a festival?

$$\begin{array}{r} 2\ R6 \\ 9\overline{)\ 24} \\ -\ 18 \\ \hline 6 \end{array}$$

They will need **3 vans.** Two vans can carry a total of 18 people. A third van is needed for the 6 remaining passengers.

B. Solve the problems. Decide what to do with the remainder.

3. To make a small bow for her daughter's costume, Anna needs a 7-inch piece of ribbon. How many bows can she make from a 186-inch ribbon?

4. Movie tickets cost $6 each. Will $100 be enough to buy 16 tickets?

5. A trail mix recipe makes 245 ounces. If the mix is stored in 8-ounce jars, how many jars are needed?

Making Connections: Divisibility

A number is evenly divisible by another if there is no remainder. Look at these rules to tell if a number is evenly divisible by the numbers listed.

A number is evenly divisible by

- **2** if the last digit is 2, 4, 6, 8, or 0
- **3** if the sum of the digits is a multiple of 3
- **4** if the last two digits form a number that is divisible by 4
- **5** if the last digit is 5 or 0
- **6** if the number is divisible by 2 *and* 3
- **8** if the last three digits form a number that is divisible by 8
- **9** if the sum of the digits is divisible by 9
- **10** if the last digit is zero

Use the rules above to see if each number in the first column at right is divisible by 2, 3, 4, 5, 6, 8, 9, and 10. Place an *X* in each column when a number is divisible by that number.

	2	3	4	5	6	8	9	10
73,908								
1,233								
2,640								
1,001								

Think of a number to exchange with a partner. Then test that number for divisibility.

Zeros in the Answer

To be sure you have not forgotten to put a zero in the answer, check how the numbers are lined up.

Putting Zeros in Answers

Example: 408 ÷ 4   Estimate: 400 ÷ 4 = 100

```
    12          Wrong
4 ) 408         The 2 cannot be placed above
  − 4           the zero because there are two
  ───           4s in 8, not in zero.
   08
  − 8
  ───
    0
```

```
    102         Right
4 ) 408         There must be a zero in the answer to show
  − 4           there are no 4s in zero.
  ───           An estimate will help you see if you have left
   00           out a zero.
  − 0
  ───
   08
  − 8
  ───
    0
```

A. Divide the problems below. Watch out for the zeros.

1. 3) 306   8) 309   5) 5,143   9) 10,008

2. 7) 56,000   9) 1,809   8) 64,080   4) 3,600

When you are working with zeros, you might want to use a short form of division. Look at the example below.

Using the Short Form

Example: 42,014 ÷ 7   Estimate: 42,000 ÷ 7 = 6,000

Standard Form

```
      6,002
7 ) 42,014
  − 42
  ─────
     00
    − 0
  ─────
     01
    − 0
  ─────
     14
   − 14
  ─────
      0
```

You can skip multiplying by zero and just bring down the next digit.

Short Form

```
      6,002
7 ) 42,014
  − 42
  ─────
    014
  − 14
  ─────
      0
```

You may bring three digits down together. Remember to put a zero above each digit before the nonzero digit.

Your answer of **6,002** is close to the estimate of 6,000.

B. Use the short form to divide the problems below.

3. $6\overline{)32,412}$ $5\overline{)651,035}$ $2\overline{)130,042}$

4. $9\overline{)45,027}$ $3\overline{)80,004}$ $8\overline{)12,120}$

Dividing by Multiples of 10

When you divide by multiples of 10, you can simplify by canceling common zeros.
First cancel the same number of zeros in the divisor and dividend. Then divide the
remaining digits as usual.

Examples:

$400 \div 100 = \frac{4\cancel{00}}{1\cancel{00}} = \frac{4}{1} = 4$ $6,000 \div 200 = 30$
          Cancel two zeros from both numbers.

C. Divide the problems below. Cancel zeros if possible.

5. $570 \div 10$ $6,200 \div 100$ $75,000 \div 100$ $90 \div 10$

6. $28,000 \div 70$ $604,000 \div 2,000$ $35,600 \div 400$ $95,000 \div 500$

Making Connections: Estimating Division

To estimate in division, round to numbers based on the division facts and cancel an
equal number of zeros. Then divide the remaining digits.

Examples: Estimate $578 \div 34$ Estimate $25,104 \div 892$
          Round to $600 \div 30$ Round to $27,000 \div 900$

$\frac{60\cancel{0}}{3\cancel{0}} = 20$ $\frac{27,0\cancel{00}}{9\cancel{00}} = 30$

Estimate each answer below.

1. $143 \div 75$ $6,524 \div 62$ $3,710 \div 58$

2. $9,876 \div 34$ $4,925 \div 79$ $23,198 \div 605$

3. Hector will receive his $18,745 lottery prize in monthly payments. Estimate the monthly
   payment if he gets his money over

   a. 36 months

   b. 48 months

   c. 60 months

Finding an Average

The **average,** or **mean,** of a group of numbers is the number that represents the typical value of the whole group.

▶ To find the average of a group of numbers
Step 1. Find the total of the numbers.
Step 2. Divide by the number of numbers.

Finding an Average

Example: If you read 15 pages on Monday, 22 pages on Tuesday, and 14 pages on Wednesday, what is the average number of pages you read per day?

Step 1. Add to find the total. 15 + 22 + 14 = 51

Step 2. Divide by the number of days.
$$3\overline{)51} \quad 17$$

You read an average of **17 pages per day.**

A. Solve the problems below.

1. The attendance at the summer conference was 9,456 people, 7,968 people, and 4,500 people during the first 3 days. What is the average attendance for those 3 days?

2. The halfway point between two numbers is the average of the numbers. Find the number halfway between 16 and 24 on the number line.

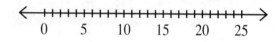

```
0    5   10   15   20   25
```

B. At the ski resort, the temperature is important. Ideal snow conditions occur when the average temperature is no more than 2° above 32°. (*Note:* The symbol for degrees is °.) Use the chart below to answer the following questions.

3. Find the week's average midday temperature. On average, were snow conditions ideal at noon that week?

4. What was the average daily snowfall from Monday through Friday?

5. What was the temperature halfway between the highest and lowest midday temperatures that week?

6. Were the snow conditions at noon on Friday and Saturday ideal? Why or why not?

Day	Temperature at Noon	Inches of Snow
Sunday	30°	6"
Monday	18°	2"
Tuesday	22°	4"
Wednesday	29°	5"
Thursday	20°	1"
Friday	36°	3"
Saturday	34°	0"

C. A recent survey shows a shopping center boom. The new stores shown
   on the graph will anchor centers built over the next three years.
   Use the graph to answer the questions.

New Shopping Centers

7. Based on the graph, estimate the average
   size of a new anchor store.

8. Find the square footage that is halfway
   between the largest and smallest stores
   shown. How many stores on the graph
   are smaller than the halfway point?

9. Suppose you use a million square feet
   of land to build a shopping center with a
   Super KMart as its anchor store plus 16
   other stores. What is the average size of
   the other stores?

D. To find how many hits a baseball player would average for 1,000 at bats,
   first multiply the number of *hits* by 1,000. Then divide by the number
   of times at bat. (Disregard any remainders.) Find the number of hits for
   1,000 at bats for the players below.

Milwaukee Brewers			
Player	**At Bats**	**Hits**	**1,000 at Bats**
Vaughn	70	20	20 × 1,000 = 20,000 20,000 ÷ 70 = _____
Spiers	260	90	_____
Diaz	410	90	_____
Jaha	320	80	_____

Discuss these questions with a partner.

10. **a.** Which player do you think would get the most hits in 1,000 at bats?

    b. the fewest hits?

11. Would you have predicted the answers to problem 10 by looking at the statistics
    provided in the chart above?

12. How can players improve their batting averages?

Mixed Review

A. Solve the problems below by division.

1. $7 \div 7 =$ $6 \div 6 =$ $30 \div 5 =$ $0 \div 4 =$ $9 \div 1 =$

2. $18 \div 6 =$ $56 \div 7 =$ $12 \div 3 =$ $20 \div 5 =$ $24 \div 8 =$

3. $15 \div 5 =$ $54 \div 9 =$ $36 \div 4 =$ $24 \div 6 =$ $54 \div 6 =$

4. $21 \div 3 =$ $18 \div 3 =$ $27 \div 9 =$ $25 \div 5 =$ $12 \div 2 =$

5. $42 \div 7 =$ $5 \div 1 =$ $35 \div 7 =$ $18 \div 9 =$ $45 \div 5 =$

6. $72 \div 8 =$ $0 \div 2 =$ $22 \div 2 =$ $30 \div 6 =$ $56 \div 8 =$

B. Divide to solve these problems. Estimate first.

7. $5\overline{)55}$ $8\overline{)96}$ $5\overline{)645}$ $2\overline{)3,816}$ $9\overline{)657}$

8. $6\overline{)4,830}$ $7\overline{)2,142}$ $3\overline{)6,509}$ $4\overline{)19,608}$ $6\overline{)6,036}$

9. $7\overline{)33}$ $3\overline{)563}$ $6\overline{)42,304}$ $5\overline{)6,342}$ $8\overline{)10,066}$

10. $240 \div 10 =$ $7800 \div 100 =$ $38,400 \div 400 =$

11. $20\overline{)840}$ $50\overline{)4,500}$ $80\overline{)45,360}$ $30\overline{)6,030}$

12. $4,210 \div 90 =$ $22,300 \div 500 =$ $5,670 \div 70 =$

C. Use your division skills to solve the following problems.

13. If you and 5 friends contribute a total of $756 to the Salvation Army Children's Fund, what is the average contribution per person?

14. Find the value of the variable in each problem below.

$\frac{32}{8} = N$ $\frac{560}{t} = 56$ $\frac{z}{7} = 7$ $\frac{b}{b} = 1$

$N =$ $t =$ $z =$ $b =$

15. Check the answers to the problems below. If they are not correct, find the right answer.

$$\begin{array}{r} 77\ R7 \\ 5\overline{)387} \end{array} \qquad \begin{array}{r} 473 \\ 9\overline{)4,257} \end{array} \qquad \begin{array}{r} 86\ R3 \\ 6\overline{)4,839} \end{array}$$

16. Charlotte filled up her car with gas. Her car can usually travel at least 300 miles on one tank of gas.

   a. If the gas tank holds about 10 gallons, what kind of gas mileage (miles per gallon) does the car get?

   b. Charlotte travels a total of 70 miles to and from work each day. How many trips to and from work can she make on one tank of gas?

17. The Reeds have a small orchard they need to fence. They plan to put fence posts every 5 yards.

   a. They already have the corner post in place. If they have to put up fence for 200 yards, how many posts do they need?

   b. If posts come in bundles of 12, how many bundles do they need to buy?

18. Ten employees purchased a Lotto ticket together. When their number was drawn, they won $12,000,000 to be paid in 20 equal payments over 20 years. If they share the winnings equally, how much will each employee receive per year?

D. With a partner discuss each problem below. Use the numbers 2, 3, 4, 5, 6, 7, 8, 9, and 10 as divisors. You can refer back to page 143 for ideas if needed.

Choose a divisor that will divide evenly in each problem. There should be no remainders. Use each number only once and use all numbers.

19. $\overline{)363}$ $\overline{)2,512}$ $\overline{)5,760}$

20. $\overline{)1,827}$ $\overline{)6,345}$ $\overline{)462}$

21. $\overline{)34,006}$ $\overline{)77}$ $\overline{)59,168}$

22. Choose one of the problems in numbers 19–21 above and make up a real-life situation to go with it.

Dividing by Two or More Digits

When you divide by two or more digits, you use both educated guessing and estimation to help you find the answer. If your first guess is not close enough, try a new number.

Dividing by a Two-Digit Number

Example: 2,080 ÷ 32   Estimate: 2,100 ÷ 30 = 70

Step 1
32 won't go into 2 or 20.
Estimate how many times 32 goes
into 208.
Ask, "How many 3s are in 20?"
Try 6.
Put the 6 above the 8 in 208.

$$\begin{array}{r} 6 \\ 32\overline{)2,080} \end{array}$$

Step 2
Multiply: 32 × 6 = 192
Subtract. Bring down the zero.
Ask, "How many 3s are in 16?"
Try 5.

$$\begin{array}{r} 6 \\ 32\overline{)2,080} \\ -1\,92 \\ \hline 160 \end{array}$$

Step 3
Multiply: 32 × 5 = 160
Subtract.

$$\begin{array}{r} 65 \\ 32\overline{)2,080} \\ -1\,92 \\ \hline 160 \\ -160 \\ \hline 0 \end{array}$$

Your answer of **65** is close to the estimate of 70.

A. Divide by these two-digit numbers.

1. $35\overline{)70}$ $24\overline{)360}$ $56\overline{)672}$ $48\overline{)960}$ $64\overline{)808}$

2. $18\overline{)378}$ $25\overline{)475}$ $34\overline{)148}$ $58\overline{)600}$ $94\overline{)8,072}$

3. $75\overline{)16,250}$ $26\overline{)16,718}$ $56\overline{)117,072}$ $63\overline{)73,773}$

B. Divide by these three-digit numbers.

4. 15,600 ÷ 325   13,728 ÷ 624   19,188 ÷ 369   122,310 ÷ 302

5. 118,887 ÷ 295   108,720 ÷ 120   11,448 ÷ 106   89,694 ÷ 297

C. Practice your division skills with this cross number puzzle.
(*Hint:* First fill in answers you can do quickly; see 18 Down.)

Across

1. 225 ÷ 9

3. 97,770 ÷ 30

7. $\frac{75}{5}$

9. 328 ÷ 4

10. $4\overline{)16,020}$

11. 140 ÷ 7

12. 966 ÷ 7

14. $6\overline{)4,254}$

15. 496 ÷ 8

16. $5\overline{)200}$

17. 760,000 ÷ 200

19. $2\overline{)122,380}$

23. $\frac{108,886}{2}$

25. 54,439 ÷ 7

28. 2,368 ÷ 37

30. 800 ÷ 32

31. 53,816 ÷ 62

34. $4\overline{)3,648}$

36. $\frac{900}{90}$

37. $48\overline{)192,720}$

39. 664 ÷ 8

40. $\frac{3,000}{150}$

41. 54,000 ÷ 9

42. 8,736 ÷ 112

Down

1. 843 ÷ 3

2. 10,468 ÷ 2

3. $\frac{68}{2}$

4. $9\overline{)1,863}$

5. $\frac{1,000}{2}$

6. $5\overline{)4,796,520}$

7. 11,340 ÷ 9

8. 20,080 ÷ 4

13. $3\overline{)2,418}$

18. $\frac{56}{7}$

20. $5\overline{)789,230}$

21. 98 ÷ 7

22. $\frac{940}{10}$

24. $75\overline{)27,675}$

25. 57,696 ÷ 8

26. $2\overline{)15,000}$

27. $\frac{56}{8}$

29. $60\overline{)251,220}$

32. $\frac{18,000}{30}$

33. $6\overline{)4,860}$

35. 4,046 ÷ 17

38. 5,000 ÷ 100

Item Sets

Sometimes math problems are given as **item sets.** An item set is a lot of information given in a paragraph, a chart, or a picture. Many questions can be asked about the information.

The key to solving problems based on item sets is to *choose only the information needed* to answer that particular question.

A. Solve problems 1–3 using the following information.

Only 5 employees work at Aunt Betty's Bagels. This chart shows the hours worked and wages paid to each employee.

Aunt Betty's Bagels Payroll		
Name	Hours	Weekly Wages
Kimly	40	$320
Leon	48	$480
Bernie	60	$900
Shirley	50	$350
Allan	25	$150

1. What is the average weekly wage paid at Aunt Betty's Bagels?

   Information needed: _____

   Calculation: _____

2. How much does Bernie make per hour?

   Information needed: _____

   Calculation: _____

3. *Write a problem.* Make up a division question using the information in the chart. Then solve your own problem.

B. Answer problems 4–7 from the following information.

Tomas keeps a mileage log of his travel as a salesman. Each time he stops for gas, he records the number of miles he has traveled.

	Mileage Log			
Day	Beginning Odometer	Ending Odometer	Miles Driven	Gallons of Gas
Monday	21,345	21,555		7
Tuesday	21,555	21,735		6
Wednesday	21,735	21,855		4
Thursday	21,855	22,095		8
Friday	22,095	22,185		3
				Total:

4. Find the number of miles Tomas has driven each day.

5. What is Tomas's total mileage for the week?

6. To find the number of miles driven per gallon of gasoline, divide the total miles by the total gallons of gas. What is Tomas's gas mileage? Do you think his gas mileage is good?

7. What is the average number of miles that Tomas drives per day?

C. Nutrition information is printed on the package of food products. Food experts say that no more than 30 percent of the calories you eat should come from fat.

You can check a product for a percentage of fat by following these steps:

1. Multiply the grams of fat by 900.

2. Divide the result by the number of calories.

3. Is your answer less than 30? If so, less than 30 percent of its calories come from fat.

Look at the labels below. Find the percentage of fat. Is it more or less than 30 percent?
Note: % means "percent."

8.

Ripe Olives

Nutrition Facts

Serving Size 5 olives (15 g)
Servings Per Container About 11

Amount Per Serving

Calories 30 Calories from Fat 27

% Daily Value

Total Fat 3g	5%
Monounsaturated Fat 1.5g	
Cholesterol 0mg	0%
Sodium 115mg	5%
Total Carbohydrate 1g	0%
Protein 0g	0%

9.

Complete Pancake Mix

Nutrition Facts

Serving Size 4 four-inch pancakes
Servings Per Container 10

Amount Per Serving

Calories 230 Calories from Fat 50

% Daily Value

Total Fat 6g	9%
Saturated Fat 1.5g	8%
Cholesterol 70mg	24%
Sodium 670mg	28%
Potassium 155mg	4%
Total Carbohydrate 37g	12%
Dietary Fiber 1g	5%
Sugars 8g	
Protein 8g	

10.

Peanut Butter Filled Crackers

Nutrition Facts

Serving Size 6 cracker sandwiches
Servings Per Container 12

Amount Per Serving

Calories 210 Calories from Fat 90

% Daily Value

Total Fat 10g	15%
Saturated Fat 3g	11%
Cholesterol 0mg	0%
Sodium 430mg	18%
Total Carbohydrate 23g	8%
Dietary Fiber 1g	4%
Sugars 5g	
Protein 5g	

Dividing Dollars and Cents

Here are some tips to help you when you divide dollars and cents.

Tip 1. If a dollar amount is given without cents, add a decimal point and 2 zeros.

    Example: $14 = $14.00

Tip 2. If money is given with a cents sign, you may rewrite it with a dollar sign and decimal point.

    Example: 34¢ = $.34

Tip 3. Divide the numbers as usual, carefully placing the digits of the answer in the proper places. Write the decimal point in the answer straight above the decimal point in the problem.

    Example: $12.42 ÷ 9

$$\begin{array}{r} \$1.38 \\ 9\overline{)\$12.42} \end{array}$$

Tip 4. There may be a remainder when you divide money.

    Example: Divide $3 evenly among 8 children.

$$\begin{array}{r} \$0.37 \ R\$.04 \\ 8\overline{)\$3.00} \end{array}$$

Each child gets 37¢. There is 4¢ left over.

Tip 5. After the decimal point in the answer, both decimal places must be filled in. Put a zero above the digit if it cannot be divided.

    Example: $1.44 ÷ 72

$$\begin{array}{r} \$.02 \\ 72\overline{)\$1.44} \end{array}$$

A. Divide the amounts below.

1. $5.75 ÷ 5 $52.47 ÷ 9 $86.52 ÷ 12

2. $926.50 ÷ 25 $3,255 ÷ 15 $28 ÷ 56

3. $1.28 ÷ 32   54¢ ÷ 9 $.72 ÷ 4

4. $6 ÷ 75 $6.90 ÷ 15 $48 ÷ 10

B. For more practice divide the money below.

5. $.64 ÷ 4 $24.15 ÷ 7 $2,502.36 ÷ 3

6. $970 ÷ 20 $2,449.60 ÷ 40 $1,400 ÷ 28

7. $4.55 ÷ 65 $34.25 ÷ 25 $7.40 ÷ 148

C. Use your division skills to solve the following. Estimate first.

8. A friend paid $552 to take 16 credit hours at the local community college. How much did he pay per credit hour?

9. Ali needs 4 new tires for her car. According to this sign, how much will she actually pay for each tire?

> **Tire Sale**
> Buy 3 tires at $48.72 each and get the fourth tire free!

10. Joe and his 2 sisters plan to split the cost of their parents' anniversary party evenly. Their purchases are listed below. Find each person's share of the cost.

Flowers	$48.75
Cake	$24.36
Decorations	$36.48
Invitations	$19.69
Food	$184.39
Beverages	$42.53
Film	$12.98

D. The local travel agency used phone interviews to survey 8 families about their vacation plans. Use the results of the survey to answer problems 11–14.

Survey questions:

- How many days long is your vacation?
- How many people in your family will go on the vacation?
- How far away is your destination?
- What is a reasonable price for a hotel room?
- How much will you spend per person each day for food?
- How much will you spend on souvenirs?

Family	A	B	C	D	E	F	G	H
Days	7	7	4	14	10	12	4	14
People	2	2	5	4	3	6	6	4
Miles	250	175	84	320	1,026	290	95	848
Hotel	$58	$42	$70	$48	$50	$75	$65	$80
Food	$45	$22	$20	$25	$30	$32	$20	$30
Souvenirs	$200	$40	$150	$80	$75	$145	$90	$100

11. What is the average number of miles these families will travel on vacation?

12. Compare the amount family G will spend per person on souvenirs with the amount family H will spend per person.

13. What is the total amount family D will spend on food and lodging for one day? What is the actual cost per person?

14. **Write** Think of another question using the information in this survey that might be helpful to a travel agency.

Dividing on a Calculator

Follow these steps when dividing on a calculator:

- Press On/Clear.
- Enter the dividend (the number being divided) *first*.
- Press the division key ÷ .
- Enter the divisor.
- Press the equal sign = .

Be sure to estimate your answer to check your work. Always ask yourself, "Does my answer make sense?"

Example: 188,928 ÷ 576

Key in:	Your display reads:	Estimate
On/C	0.	$$\overset{30\ 0}{6\emptyset\emptyset\,\overline{)180,000}}$$
1 8 8 9 2 8	188928.	
÷	188928.	*or*
5 7 6	576.	$\frac{180,0\emptyset\emptyset}{6\emptyset\emptyset} = 300$
=	328.	

Practice this procedure on your calculator.

Follow these tips when you use a calculator for division:

- Be careful to *enter the divisor after* the division sign.
- Use the decimal point when you enter dollars and cents.
- Remainders from division are shown as decimals on the calculator.

Example: 25 ÷ 4 = 6.25 on the calculator.

To find the remainder, multiply the divisor by the whole number part of the answer. Then subtract from the original number being divided.

whole number answer ⌐
          ↓
4 × 6 = 24   25 − 24 = 1
25 ÷ 4 = 6 R1   └─ remainder

A. Identify the divisor in each problem. Then use your calculator to find the answer.

1. 184 ÷ 92 $16\overline{)1,680}$ $\frac{3,075}{25}$

Divisor: _____   Divisor: _____   Divisor: _____

Answer: _____   Answer: _____   Answer: _____

2. How many times does 48 go into 154,320?

Divisor: _____

Answer: _____

3. Find the quotient of 22,515 divided by 395.

Divisor: _____

Answer: _____

4. How many 9s are in 25,704?

Divisor: _____

Answer: _____

B. Use a calculator to solve the problems below. Estimate first.

5. $26\overline{)12{,}610}$ $26{,}520 \div 78$ $\frac{11{,}655}{45}$

Estimate: _____   Estimate: _____   Estimate: _____

Answer: _____   Answer: _____   Answer: _____

6. $75.24 \div 3$ $218 \div 4$ $3{,}690.45 \div 15$

Estimate: _____   Estimate: _____   Estimate: _____

Answer: _____   Answer: _____   Answer: _____

C. The answers below have remainders. Use your calculator to divide and then find the remainder.

7. $476 \div 9$ $8{,}095 \div 12$ $24{,}315 \div 72$

Making Connections: Cutting the Cake

One part of the catering business is serving a special-occasion cake. It might be a birthday or wedding cake. The caterer can use a calculator and the divisibility rules to find all the different ways to serve the cake.

Suppose there are 32 guests. The cake could be cut as shown.

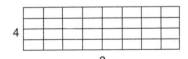
4

8

or

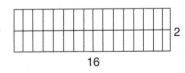
2

16

Show how the caterer might cut the cake if he has

20 guests

35 guests

60 guests

150 guests

Choosing the Right Operation

There are four arithmetic operations: addition, subtraction, multiplication, and division. When you solve a problem, look for ideas that will help you decide which operation to use.

Use this chart to help you recall what ideas each operation brings to mind.

▶ Addition	▶ Subtraction	▶ Multiplication	▶ Division
Idea: *combine* numbers to get a larger amount	**Idea:** *find the difference* between two numbers for comparison **Idea:** *take away* an amount to make the value smaller	**Idea:** *combine like amounts* to get a total	**Idea:** *separate* into equal parts **Idea:** find *how many times* one amount fits into another

A. Discuss this chart with a partner. Give an example of a time in your daily life that you could use each operation.

B. Read the problems below. Choose the operations you would use to solve the problem. *Do not do the calculations.*

1. While doing laundry, Corey found two $10 bills, a single, two quarters, and a dime that she had washed. How much did she find altogether?

   Operation: _____

2. To run for election, you need 500 registered voters to sign a petition. There are 9,500 registered voters out of a town population of 18,340 people. How many candidates could run for election if each voter is allowed to sign only one petition?

   Operation: _____

3. How much do you save off the full price if you buy a wool suit and a pair of cotton pants?

   Operation: _____

Designer
Wool Suits
Regular and
Athletic Fit
**$149**^{95}
(Compare at $325)

Wrinkle Free
Casual Pants
100% Cotton
**$19**^{95}
(Compare at $38)

C. **Carl and Roberta have decided to redecorate their family room. The floor plan is shown below. Help them figure their expenses.**

4. A gallon of paint covers 400 square feet of wall space. Roberta found the surface area of her 8-foot walls to be approximately 672 square feet (not counting doors and windows). If she wants 2 coats of paint on the walls, how many gallons of paint does she need to buy?

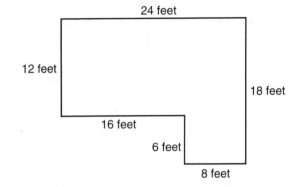

5. Carl looked at several brands of carpeting. The ones he liked best were $12.95 per square yard, $18.95 per square yard, $16.50 per square yard, and $15 per square yard. What is the average price of his choices?

7. If 1 square yard equals 9 square feet, how many square yards of carpeting must Carl buy to cover his floors?

6. How many square feet of carpeting is required to cover the family room floor?

8. Roberta chose paint on sale for $9.99 per gallon, and Carl selected carpeting for $16.50 per square yard. What are their total expenses?

D. **The chart below shows how many calories you can burn in 15 minutes. Use this information to answer the questions.**

9. How many calories do you burn on a 2-hour fitness walk?

10. A triathlete swims, runs, and bikes in competition. What is the average number of calories burned every 15 minutes if the athlete swims 15 minutes, bikes 1 hour, and jogs for half an hour?

11. **Multiple Solutions** Select a combination of at least two activities that burns more than 400 calories total in an hour. What combination will burn the most calories in 1 hour?

How Much Will I Burn?	
The following chart shows how many calories you can burn in just 15 minutes.	
Activity	**Calories Burned**
Walking (.5 miles)	83
Fitness Walking (1.1 miles)	100
Jogging (1.5 miles)	160
Running (1.8 miles)	180
Cycling (2.4 miles)	102
Cross Country Skiing	190
Swimming (600 yards)	110
Handball/Racquetball	150
Aerobic Dancing	105

Finding Unit Price

Bran Cereal	K-Pops	Oat-O's	Wheat Treats
14 oz. $1.96	19 oz. $2.85	20 oz. $3.20	18 oz. $2.34

Which cereal is the best buy in terms of cost? The labels tell you the boxes are measured in ounces (oz.). If you know the price for 1 ounce, you will be able to choose the best buy.

► To find the **unit price** (price for one unit), divide the total cost by the number of units.

Example: Bran Cereal costs $1.96 for 14 ounces.

$$14\overline{)\$1.96}\quad \$0.14$$

Bran Cereal costs **14¢ per ounce.**

A. Find the unit price for the other three boxes of cereal at the top of the page. Which cereal has the lowest unit price?

1. K-Pops: **2.** Oat-Os: **3.** Wheat Treats:

Sometimes when you divide to find unit price, there is a remainder. If so, follow the usual store pricing rule: round up to the next penny.

B. Find the unit price of the items below. Some problems require more than one step.

4. Buy 3 cassette tapes for $8.96 each. Get the 4th tape free.

Unit price: _____
per cassette

6. Eight bottles of cola

$3.28

Unit price: _____
per bottle

5. Buy 5 lb., get 5 lb. free

5 lb. for $1.50

Unit price: _____
per pound

7. Shoe sale

1st pair $22.98
2nd pair half price

Unit price: _____
per pair of
shoes

C. Kim Lee owns a company that supplies team uniforms. He buys items in bulk and sells them by the piece. To determine the selling price, he finds the unit price and adds an amount for profit. You may use a calculator.

8. For each item below, find the unit price. Then double that price to find the selling price. (A case is 24 pieces.)

Inventory

Item	Cost	Unit Price	Selling Price
Quality T-shirt	$49.92/doz.		
Long sleeve T-shirt	$93.60/doz.		
Baseball caps	$156/case		
Golf shirts/pocket	$57.90/6-pack		
Heavyweight sweatshirt	$237.80/doz.		
Jersey shorts	$49.80/6-pack		

9. If Kim Lee sells one of each item at his selling price, what is his average price per item?

10. **Describe** Have you ever worked at a store? Explain how the selling price of the merchandise was determined.

D. Your ability to find unit price may help you get the best deal when you shop. At the grocery store, unit price is often posted. Select the best deal after you calculate the unit prices below.

11.

Dog Food 40 pounds
$17.99

Dog Food 8 pounds
$5.99

Unit price: _____
per pound

Unit price: _____
per pound

13.

strawberries

12 pints
$5.98

1 pint
79¢

Unit price: _____
per pint

Unit price: _____
per pint

12.

$2.89
1 gal.

$.89
1 qt.

$1.64
$\frac{1}{2}$ gal.

Unit price: _____ _____ _____
per quart   per quart   per quart

Note: 1 gallon = 4 quarts

14. **Discuss** Discuss your answers with a partner. Is the lowest unit price always a good deal for you? What other things would you need to consider when shopping besides unit price?

Putting It All Together

There are many problems that you can solve with the skills you have developed. Addition, subtraction, multiplication, and division are the foundation of mathematics.

Use your skills often. If you combine your ability to estimate with the use of a calculator, you will have a lot of success in mathematics.

So get out your calculator and let's practice.

A. Assume your grocery bills for the 4 weeks of February are listed below. First estimate your average weekly grocery expense. Then use your calculator to find the exact amount.

$90.81, $110.12, $128.60, $105.23

Estimate: _____   Exact: _____

B. You want to make sure your paycheck is correct. You make $8.72 per hour for regular time and $13.08 for overtime. Last week you worked 40 hours plus 12 hours overtime. Estimate your pay and then find the exact answer.

Estimate: _____   Exact: _____

C. You've decided to tile the kitchen floor shown below. If each floor tile is 1 square foot, find the number of tiles you will need. In addition, find the number of feet of baseboard you will need to trim around the room.

Number of tiles: _____

Length of baseboard: _____ feet

```
                          |— —|
                          3 ft.
                          door
          Kitchen              9 feet

          16 feet
```

D. Estimate. How much do you think it costs to own a car? Fill in your own numbers. Then find average yearly and monthly costs.

Monthly car payment: _____

One year of insurance: _____

One year of gas and oil: _____

Miscellaneous repairs: _____

162

E. **Different methods have been used to compute welfare grants over the years. In the mid-1990s, welfare grants were calculated as follows. With a partner, discuss the information and these questions.**

A social worker calculates the amount of money a welfare recipient gets. The maximum grant in this region for a mother with 2 children is $724 per month. If the mother has a job, the amount of the grant is reduced. Assume a welfare recipient with 2 children earns $750 a month at her job. Follow these steps for reducing the grant:

Step 1. From the amount she earns at work, subtract $75 for work-related expenses.

Step 2. Subtract $150 per child for child care.

Step 3. Subtract $30.

Step 4. Divide what remains by 3 and subtract that amount from the amount you found in step 3.

Step 5. Subtract this amount from the grant.

1. Follow steps 1 and 2 above. What amount did you get? This is what the government expects will be left of her wages after the worker pays for child care and all other work-related expenses.

2. Continue with steps 3–5 above to find the amount of the reduced grant.

3. Combine the answers to problems 1 and 2. This gives the total amount the worker gets from the reduced grant and what's left of her wages after she pays work-related expenses.

4. Discuss with a partner or the class the advantages and disadvantages of this recipient finding employment.

5. **Investigate** Find the method for calculating welfare grants in your area. Compare your findings with the method shown above. Discuss.

F. **Just for fun, find the secret number *n* from the clues below.**

- $n < 100$
- n is even.
- $n > 7 \times 8$
- The sum of the digits is 12.
- n is evenly divisible by 7.

What's the number? _____

Make up your own secret number and write the clues.

Unit 5 Review

A. Fill in each blank to make each statement true.

1. 5,789 rounded to the nearest thousand is _____.

2. 312 > _____

3. If it's 7:15 now, then in $3\frac{1}{2}$ hours it will be _____.

4. If $T + 12 = 20$, then $T = $ _____.

5. 253 + 60 + 4 = _____

6. _____ is the perimeter of a square 8 inches on each side.

7. The next number in the pattern 8, 5, 10, 7, 12, is _____.

8. 45¢ plus 75¢ is _____.

9. $18 + $9.65 = _____

10. If $m - 38 = 10$, then $m = $ _____.

11. 3,000 − 856 = _____

12. If $R - R = 0$, then $R = $ _____.

13. If $F + 1 = 75$, then $F = $ _____.

14. _____ is the area of a rectangle 9 feet by 8 feet.

15. 34 × 100 = _____

16. If $7n = 63$, then $n = $ _____.

17. 96 × 30 = _____

B. Use your division skills to solve the problems below.

18. 56 ÷ 8 =   45 ÷ 9 =   18 ÷ 6 =   42 ÷ 7 =   60 ÷ 10 =

19. 27 ÷ 3 =   64 ÷ 8 =   63 ÷ 9 =   28 ÷ 7 =   81 ÷ 9 =

20. $\frac{250}{10}$ $\frac{4,800}{20}$ $\frac{560}{70}$ $\frac{900}{90}$ $\frac{74,000}{100}$

21. $5\overline{)5,245}$ $7\overline{)296}$ $15\overline{)345}$ $68\overline{)4,725}$ $25\overline{)1,000}$

22. $80\overline{)9,430}$ $32\overline{)64,032}$ $205\overline{)4,305}$ $99\overline{)500}$

C. Solve the problems below.

23. Wesley plans to edge his garden with 8-inch bricks. Use the drawing below to find how many bricks he will need.

12 ft.

8 ft.

24. The local bakery sold 1,308 donuts last Friday. If the donuts were sold for $4.65 per dozen, how much money was made from the sales?

25. Test the divisibility of 53,640. What numbers between 1 and 10 is it evenly divisible by?

26. Find the value of each variable.

$$\frac{H}{10} = 42 \qquad \frac{W}{W} = 1 \qquad \frac{x}{8} = 8$$
$$H = \qquad\qquad W = \qquad\qquad x =$$

D. The local library is filled with books, magazines, cassette tapes, and videotapes for residents to borrow. Solve and discuss the problems below with a partner.

27. You can borrow a book for 14 days. If a book has 294 pages, how many pages must you read per day to return the book on time?

28. The videotape collection includes 64 classics, 48 children's features, 28 educational tapes, and 30 current movies. The value of the collection is $4,080. What is the average cost of a videotape in this collection?

29. By the end of the week, a library had collected a total of $27.75 on 37 late books. If that week's average holds for all books, what can the library expect to collect on 50 late books?

30. Discuss Where is your local library? Discuss with a partner how you can use your library for both education and relaxation.

Working Together

The newspaper is a great source of information. Investigate the following areas with a partner or small group.

Locate the listing of temperatures across the country. Find the average temperature of 10 cities in the region of your choice.

In the classified section, find and list different used car prices for similar cars. What factors do you think could affect used car prices?

Find the average of the prices you listed. Do you think the average gives you a good idea of how much a similar used car would cost?

Solve the following problems.

1. If you travel the entire Pennsylvania Turnpike, you must pay 9 tolls. Find the total cost at 60¢ per toll.

   (1) 60¢

   (2) 90¢

   (3) $3.60

   (4) $5.40

   (5) $54

Problem 2 is based on the order form below.

Hansen's Wood Products			
Qty	Item	Each	Total
1	round table	$596	
4	chairs	$108	
1	hutch	$875	
		Subtotal:	
		Tax:	$95
		Delivery:	$50
		Total:	

2. Select the best estimate for the total cost of buying the furniture shown on the order form above.

   (1) $1,600   (4) $2,050

   (2) $1,750   (5) $2,500

   (3) $1,900

3. Sarah earns $36,000 per year. Her deductions for Social Security, federal tax, and state tax total $10,000 for the year. She has no other deductions. If she gets paid every other week, how much money does she take home per paycheck?

   (1) $500   (4) $1,769

   (2) $885   (5) $3,000

   (3) $1,000

4. Gina used a $100 bill to pay for $14.80 in gas. How much change should she get back?

   (1) $85.20   (4) $86.80

   (2) $85.80   (5) $96.80

   (3) $86.20

Problems 5 and 6 refer to the credit card application below.

330-36-7025	Taylor
Social Security No.	Mother's Maiden Name (Security Requirement)
Mo. 7 Day 5 Yr. 64	(708) 555-1156
Date of Birth	Home Telephone
8 Years 7 Months	Home ☐ Own ☒ Rent ☐ Other
How Long At Current Residence?	
Carpenter	$28,075
Occupation	Annual Household Income
Chris Caile	
Name	
Address	☐ Savings Account
( )	☐ Checking Account
Business Telephone	Do You Have The Above?
X	
Full Signature	Date

5. What is the birth date of Chris Caile?

   (1) August 5, 1964

   (2) May 7, 1964

   (3) June 4, 1975

   (4) April 6, 1975

   (5) July 5, 1964

6. What is Chris's yearly income?

   (1) twenty thousand eight hundred seventy-five dollars

   (2) twenty-eight thousand seventy-five dollars

   (3) twenty-eight thousand seven hundred five dollars

   (4) two thousand eight hundred seventy-five dollars

   (5) two hundred eighty thousand seventy-five dollars

Problem 7 refers to the advertisement below.

FAMILY PAC SAVER

- 5 lbs. Sirloin Steak • 3 lbs. T-Bone Steak
- 5 lbs. Round Steak • 4 lbs. Boneless Rump Roast
- 3 lbs. Center Cut Pork Chops
- 3 lbs. Lean Beef Stew • 3 lbs. Chuck Roast
- 6 lbs. Grade "A" Fryers • 10 lbs. Ground Chuck
- 3 lbs. Pork Loin Roast • 3 lbs. Mock Chicken Legs
- 3 lbs. Italian Sausage • 2 lbs. Sliced Bacon

BUY ONE AT **$119⁹⁹**   GET THE 2ND **$59⁵⁰**

7. What is the cost per pound (to the nearest cent) of meat if you buy 2 Family Pac Savers as shown in the ad?

   (1) $1.13   (4) $2.26
   (2) $1.69   (5) $3.37
   (3) $1.79

8. The country dancing class has 1 more woman than man attending. Which number below must be the number of people in the class?

   (1) 15   (4) 40
   (2) 22   (5) 44
   (3) 36

Problem 9 refers to the drawing below.

10 feet

17 feet

9. A carpet remnant will make a nice area rug if binding is put around the perimeter. How many feet long will the binding be when the rectangular piece is finished?

   (1) 27
   (2) 44
   (3) 54
   (4) 170
   (5) 680

Problem 10 refers to the form below.

Fashion Warehouse		Date: 9/3
Order Number	Destination	Weight
10742	New York, NY	4 lb.
15394	Washington, D.C.	10 lb.
13156	New York, NY	8 lb.
10617	Wilmington, DE	5 lb.
12784	Newport News, VA	8 lb.
17394	Charleston, NC	7 lb.
10159	Durham, NC	12 lb.
14943	Attica, NY	4 lb.
12175	Washington, D.C.	5 lb.

10. At Fashion Warehouse, Akeem packs and ships catalog orders. He also records the weight of each package. What is the average weight in pounds of the packages listed on the form?

    (1) 5
    (2) 6
    (3) 7
    (4) 8
    (5) 9

11. A new product on the market is the golf cart cooler. The factory produces 12,000 coolers per month. If all those coolers were sold and sales for the year were $576,000, what was the selling price for each cooler?

    (1) $4.00
    (2) $4.80
    (3) $5.64
    (4) $40.00
    (5) $48.00

12. A part-time trucker for Rapid Delivery Corporation makes $18.30 per hour. If she drove 42 hours total during the last two weeks, how much did she earn for that time?

    (1) $76.86
    (2) $109.80
    (3) $768.60
    (4) $1,098.00
    (5) $7,686.00

Posttest

Problem 13 is based on the advertisement below.

ASTEROID 48 Months $139 Per Month

ROMA 36 Months $189 Per Month

13. Choose the statement that best describes the comparison of the two cars shown in the ad.

(1) The Roma lease costs $50 more than the Asteroid lease.

(2) The lease for the Asteroid costs $132 more than the lease for the Roma.

(3) The lease for the Roma is $600 less than the lease for the Asteroid.

(4) The lease for the Roma costs $132 more than the lease for the Asteroid.

(5) The Roma lease is $1,668 less than the Asteroid lease.

14. Choose the answer that contains an equation to correctly match statements a–d below.

a. Two more than the number a is 12.

b. Two less than the number b is 12.

c. Twice the number c is 12.

d. The number d split into two equal parts is 12.

(1) $2a = 12$ $2 - b = 12$
   $2c = 12$ $\frac{2}{d} = 12$

(2) $a + 2 = 12$ $b - 2 = 12$
   $2c = 12$ $\frac{d}{2} = 12$

(3) $a + 2 = 12$ $2 - b = 12$
   $2c = 12$ $\frac{d}{2} = 12$

(4) $2a = 12$ $2 - b = 12$
   $2c = 12$ $\frac{2}{d} = 12$

(5) $12 + 2 = a$ $12 - 2 = b$
   $12 \div 2 = c$ $12(2) = d$

Use the following information to solve problems 15 and 16.

A typical chain letter scheme may ask you to mail the letter to 6 people. The letter requests each of those 6 people to send the letter to 6 more people.

Look at the chart below from the U.S. Postal Service that shows what happens to the chain.

No. of Mailings	No. of Participants
1	6
2	36
3	216
4	1,296
5	7,776
6	46,656
7	279,936
8	1,679,616
9	10,077,696
10	60,466,176
11	362,797,056
12	2,176,782,336
13	13,060,694,016

U.S. Population: Over 250 million

World Population: Over 5 billion

15. Which mailing needs more participants than there are people in the United States?

(1) 7
(2) 8
(3) 9
(4) 10
(5) 11

16. *About* how many times larger is the sixth mailing than the third mailing?

(1) 23
(2) 30
(3) 230
(4) 300
(5) 2,300

168

Problem 17 is based on the floor plan below.

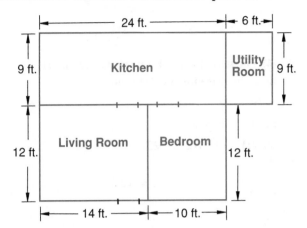

17. Find the number of square feet of linoleum needed to cover the kitchen floor and utility room floor.

    (1) 270 square feet   **(4)** 78 square feet

    (2) 216 square feet   **(5)** 48 square feet

    (3) 66 square feet

18. Deb wants to take her 4-year-old daughter to the library on Wednesday to listen to a story. According to the library schedule below, what time should she leave home if it takes her 10 minutes to get to the library and she wants to arrive 15 minutes early?

    (1) 5:35 P.M.   **(4)** 6:05 P.M.

    (2) 5:45 P.M.   **(5)** 6:35 P.M.

    (3) 5:50 P.M.

> **Toddler Tales**
> *2 1/2 - 3 1/2 year olds*
> **01** Tuesday   9:30-9:50 A.M.
> **Storytimes**
> *3 1/2 - 5 year olds*
> **02** Tuesday   10:00-10:30 A.M.
> ***** Wednesday   7:00-7:30 P.M.
> **No registration is required.*
> *This is a drop-in class.*
> **Kaleidoscope**
> *1st through third graders*
> **04** Tuesday   10:30-11:15 A.M.
> **05** Tuesday   1:00-11:45 P.M.

Popular Products posts the graph below showing sales each month. Problems 19 and 20 are based on the graph.

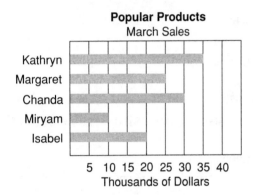

19. What is the average amount of sales per salesperson in March?

    (1) $24

    (2) $120

    (3) $20,000

    (4) $24,000

    (5) $120,000

20. Popular Products pays a $250 bonus to a salesperson for each $5,000 in sales per month. How much is the bonus for the top salesperson?

    (1) $140

    (2) $175

    (3) $1,750

    (4) $14,000

    (5) $17,500

1. **(4) $5.40**

   60¢ = $.60

   9 × $.60 = $5.40

2. **(4) $2,050**

   $600 + $400 + $900 + $100 + $50 = $2,050

3. **(3) $1,000**

   $36,000 − $10,000 = $26,000

   $26,000 ÷ 26 = $1,000

4. **(1) $85.20**

   $100.00 − $14.80 = $85.20

5. **(5) July 5, 1964**

   7/5/64

6. **(2)** twenty-eight thousand seventy-five dollars

7. **(2) $1.69**

   53 × 2 = 106 pounds

   $119.99 + $59.50 = $179.49

   $179.49 ÷ 106 = $1.69

8. **(1) 15**

   The answer must be an odd number.

9. **(3) 54**

   10 + 17 + 10 + 17 = 54

10. **(3) 7**

    4 + 10 + 8 + 5 + 8 + 7 + 12 + 4 + 5 = 63

    63 ÷ 9 = 7

11. **(1) $4.00**

    12 × 12,000 = 144,000

    576,000 ÷ 144,000 = 4

12. **(3) $768.60**

    42 × $18.30 = $768.60

13. **(4) The lease for the Roma costs $132 more than the lease for the Asteroid.**

    $139 × 48 = $6,672

    $189 × 36 = $6,804

    $6,804 − $6,672 = $132

14. **(2)** $a + 2 = 12$, $b − 2 = 12$, $2c = 12$, $\frac{d}{2} = 12$

15. **(5) 11**

    362,797,056 is more than 250 million.

16. **(3) 230**

    46,000 ÷ 200 = 230

17. **(1)** 270 square feet

18. **(5) 6:35 P.M.**

    25 minutes before 7:00

19. **(4) $24,000**

    20 + 10 + 30 + 25 + 35 = 120

    120 ÷ 5 = 24

20. **(3) $1,750**

    35,000 ÷ 5,000 = 7

    $250 × 7 = $1,750

Posttest Evaluation Chart

Make note of any problems that you answered incorrectly. Review the skill area for each of those problems, using the unit number given.

Problem Number	Skill Area	Unit
1	Multiplying money	4
2	Estimating money	2
3	Multistep	5
4	Subtracting money	3
5	Reading dates	1
6	Reading numbers	1
7	Multistep; unit price	5
8	Even and odd numbers	1
9	Finding perimeter	2
10	Multistep; average	5
11	Finding unit price	5
12	Multiplying money	4
13	Multistep; comparing prices	4
14	Equations	2–5
15	Reading numbers	1
16	Estimating	5
17	Finding area	4
18	Telling time	1
19	Average; using a data source	5
20	Multistep; using a data source	4

Answer Key

Unit 1

Everyday Numbers p. 13
Answers to the survey will vary.

Talk About It

The category for each question is shown below.

1. B	11. C
2. A	12. C
3. A	13. B
4. A	14. B
5. D	15. D
6. B	16. D
7. C	17. C
8. C	18. D
9. B	19. C
10. A	20. E

Success in Math pp. 14–15
Answers will vary.

Counting and Grouping pp. 16–17

Part A

1. 2	3. 3
2. 5	4. 1

Part B

5. 10, 20, 30, 40, 50, 60, 70, 80, 90

6. 100, 200, 300, 400, 500, 600, 700, 800, 900

7. **a.** 3, 6, 9, 12, 15, . . .

   b. 4, 8, 12, 16, 20, . . .

   c. 30, 24, 18, 12, 6

8. **T-Shirt Color Chart**

White	Gray	Black	Red	Blue	Yellow
ЖЖ	ЖЖ	ЖЖ	ЖЖ	ЖЖ	|||
ЖЖ	ЖЖ	||||	||	ЖЖ	
ЖЖ				||	
18 total	11 total	9 total	7 total	12 total	3 total

Discussion will vary.

Part C

9. 432	12. 205
10. 67	13. 70
11. 920	14. 370

Part D

15. **a.** 6	**b.** 2	**c.** 3	**d.** 4
16. **a.** 16	**b.** 4	**c.** 30, 40	

17. 8

Making Connections: Odd and Even Numbers p. 17

1. **a.** even **b.** odd **c.** odd **d.** even **e.** odd

2. **(1) 72 members**
   This is the only even number, and even numbers can be grouped into pairs.

3. **Yes,** because 15 is an odd number.

The Number Line pp. 18–19

Part A

1. 2	5. 5
2. 9	6. 5
3. 5	7. smaller
4. 8	

Part B

8. 9 > 6	11. 4 > 2
9. 15 = 15	12. 100 > 88
10. 3 < 12	

Part C

13. 3 > 1

14. 48 < 49

15. 9 = 9

Making Connections: Measuring Tools p. 19

1. 36

2. 3

3. **a.** 12   **b.** 2   **c.** 36

4. 45

5. 150

6. Answers will vary. Sample answers are speedometers, rulers, and thermostats.

Place Value pp. 20–21

Part A

1.

Places	hundreds	tens	ones
Digits	7	9	5
Values	700	90	5

seven hundred ninety-five

Places	hundreds	tens	ones
Digits	3	5	0
Values	300	50	0

three hundred fifty

Places	hundreds	tens	ones
Digits	8	5	6
Values	800	50	6

eight hundred fifty-six

2. c, b, d, a

Part B

3. 8, 80

4. 176

5. d, e, b, a, c

Part C

6. (3) 3,420,015

7. (2) 12,902

Part D

8. a. 453 < 456

   b. 72,302 > 72,032

   c. 108,343 < 180,342

9. a. 28; 82; 208; 2,008

   b. 2,345; 2,354; 2,435; 2,453

   c. 576; 765; 6,057; 60,507

Rounding pp. 22–23

Part A

	Between	Round Up or Down	Rounded Value
1.	10 < 19 < 20	up	20
2.	40 < 48 < 50	up	50
3.	70 < 71 < 80	down	70
4.	80 < 82 < 90	down	80

Part B

5. a. 100   b. 400   c. 600

Part C

6. a. 50   b. 180   c. 90   d. 670   e. 330

7. a. 4,800   b. 12,600   c. 800   d. 800   e. 600

8. a. 3,000   b. 28,000   c. 7,000   d. 13,000   e. 2,000

Making Connections: Real-Life Estimation p. 23

Estimates will vary. Any reasonable estimate is correct.

1. between $22,000 and $22,500

2. $800

Does the Answer Make Sense? pp. 24–25

Part A

1. (1) 6

2. (2) 72° F

3. (3) 250,000,000

4. (1) 55

5. (2) 80

Part B

Super Sport: $28,000 Betty

Family Sedan: $14,000 Ben

Coupe: $16,000 Mazher

Minivan: $23,000 Miguel

Part C

Explanations will vary but should be similar to the ones below.

Super Sport: $27,768 rounds to about $28,000. Betty will spend about $2,000 less than her spending limit.

Coupe: $16,394 rounds to about $16,000. Mazher can afford this amount and will have about $2,000 to spare.

Family Sedan: $14,285 rounds to about $14,000. Ben best fits this price range.

Minivan: $22,941 rounds to about $23,000. This best fits in Miguel's price range.

Part D

6. (2) exact: 2:48 P.M. You need to know exactly when a plane departs.

7. (2) exact: 32 inches. You need an exact size for an inseam.

8. (1) estimate: around 200 people. An estimate is enough for a description.

9. (1) estimate: approximately 30 feet. An estimate is enough for a description.

10. (2) exact: 43 hours. Every hour counts where your pay is concerned.

Mixed Review pp. 26–27

Part A

1. Answers will vary, but they may include your age, birthday, Social Security number, phone number, address, bank account number, and so on.

2. Answers will vary, but yours should encourage the person to overcome the math anxiety.

3. **a.** 2   **b.** 4   **c.** 1

4. **a.** 18, 21   **b.** 9, 15

5. **a.** even   **b.** odd   **c.** even   **d.** odd

6. Yes, because 76 is an even number.

7. **a.** 37   **b.** 7 units   **c.** 34

Part B

8. **a.** >   **b.** <   **c.** =

9. **a.** >   **b.** >   **c.** <

Part C

10. b, c, a, d

11. **a.** 8,732   **b.** 2,378

12. **a.** 5,000   **b.** 5,000

13. Answers will vary. The following numbers would all round to 100: 98, 88, 102, 114.

Part D

Answers will vary for the following questions. Possible answers are listed below.

14. 200; 8; 11

15. 5; 50

16. 40,000; 200; 200; 10

17. 30,000; 10; 1,000

18. Answers will vary; the rounded numbers end in zero; yes.

Calendars pp. 28–29

Part A

1. April
2. February
3. 30 days
4. 31 days

Part B

5. Wednesday
6. Sunday
7. April 17, 1999
8. **a.** 3 months
   b. 3 months
9. **a.** 14 days
   b. 23 days
10. June 26
11. August 20
12. Saturday

Using Dollars and Cents pp. 30–31

Part A

1. There are many possible answers. Sample answers are given below.

   a. $10 $5 $1 $1 $1
      10   15   16   17   18

   b. $20 $20 $20 $20 $10
      20   40   60   80   90

2. **a.** $46   **b.** $35

Part B

3. **a.** $16.51   **b.** $65.41

4. **a.** $4.00   **b.** $.04   **c.** $.40   **d.** $40.00

5. **a.** $.70   **b.** 3¢   **c.** $11.30

6. **a.** $.08, 80¢, $8, $80.00
   b. $4.08, $4.20, $4.82, $48.20

Part C

Answers will vary, but they may look like this.

7. **a.** bills: none; coins: 2 quarters
   b. bills: $20, $10; coins: none

8. $1, $1, $1, 25¢, 25¢, 25¢, 10¢, 10¢

9. Answers will vary. You can use different combinations of coins. A sample answer is 25¢, 25¢, 25¢, 10¢, 5¢

Paying with Cash pp. 32–33

Part A

1. **a.** notepad: 1¢, 1¢, 10¢, 10¢
   b. lemons: 1¢, 5¢,10¢
   c. video: 5¢, $5
   d. suit: 1¢, 1¢, 1¢, 10¢, 25¢, 25¢, 25¢, $1, $1, $1, $10

Part B

Answers will vary.

2. **a.** bills: $10 $10 $10 $10 $10 $1 $1 $1 $1
   count: 125 135 145 155 165 175 176 177 178 179
   You need $54.00. (Add the bills.)

   b. coins
   and bills: 50¢ $1 $1 $10 $1 25¢
   count: 22.50 23 24 25 35 36 36.25
   You need $13.75 (Add the coins and bills.)

c. coins

and bills: 25¢ $10 $10 $10 $10 25¢   25¢   10¢

count: 134.75 135 145 155 165 175 175.25 175.50 175.60

You need $40.85. (Add the coins and bills.)

Part C

  3. a. $7   **b.** $56   **c.** $5   **d.** $17

Part D

  4. $90

  5. $700

  6. $130

  7. $200

Making Connections: Shopping p. 33

Estimates will vary. You can round to the nearest $10 or $100. Sample estimates are given below.

  1. item: locket and earrings set; actual price: $34.99; rounded to the nearest $10: $30

  2. item: earrings; actual price: $379; rounded to the nearest $10: $380

  3. item: ring; actual price: $89; rounded to the nearest $100: $100

Checks and Money Orders pp. 34–35

Part A

  1. $28.50

  2. Margaret Lytle

  3. Corey Phillips

  4. May 12, 1994

Part B

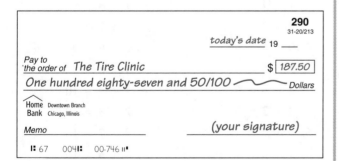

Part C

  5. $182.00

  6. Three Lakes Lumber Company

  7. Maria Flores

  8. A. M. Amirth

Part D

```
Your Currency Exchange                                    2-425
1600 W. Grand Ave.                                         710
Hometown, Il  60007                          No. A 171608
338-2055
                                          (today's date) 19____
Remitter  (your name)
                                          Dollars   Cents
Pay to
the order of  Lydia Omori, M.D.             213      00

Two hundred thirteen _____ Dollars

                                          A.M. Amirth
State regulated                  Grand Check Cashiers, Inc.
```

What Time Is It? pp. 36–37

Part A

  1. a. 3:00

      b. 7:25

      c. 10:30

Part B

Answers will vary. Sample times are given below.

  2. a. 8:15 A.M.   **b.** 6:30 P.M.   **c.** 1:00 P.M.

  3. a. 12:00 A.M.   **b.** 10:30 A.M.   **c.** 7:00 P.M.

Part C

  4. e, b, f, c, g, a, d

Tables and Schedules pp. 38–39

Part A

  1. $738 a year

  2. 35

Part B

  3. Step

  4. Monday, Wednesday, and Friday, 9–9:45 A.M.

  5. Answers will vary.

Part C

  6. 8:23 A.M.

  7. 5:05 P.M.

  8. There is no next bus; the last bus left at 5:14 P.M.

Making Connections: Daily Nutrition p. 39

  1. 3

  2. school-age children, teenage girls, adult men

  3. teenage boys, adult men

  4. 3 fruits and 4 vegetables or a total of 7

  5. Answers will vary.

Unit 1 Review pp. 40–41

Part A

1. 6,432

2. 2,346

3. 2,643 (or another number that ends with 3)

4. **a.** 2   **b.** 4

5. **a.** 32 > 23   **b.** 432 < 634   **c.** 2,436 > 2,346

6. 4,306,000

7. **a.** 3,400; rounds down because the number in the tens place (2) is less than 5

   b. 3,500; rounds up because the number in the tens place (6) is greater than 5

Part B

8. **a.** Wednesday

   b. 4; even number

   c. Thursday

   d. March 11; March 25

Part C

9. **a.** a penny. Only a penny can take 75¢ to 76¢.

   b. 3 coins: 50¢, 25¢, 1¢. These coins have the greatest value without going over 76¢.

   c. 76 pennies. Pennies are worth the least amount, so you would need more of them.

10. **a.** 9:20 A.M.

    b. 7:30 P.M.

11. c, a, e, b, f, d, g

12. **a.** $2.05; bills and coins: $1, $1, 5¢

    b. $18.46

13. Estimates will vary.

    a. about $20

    b. about $2,200

Part D

14. **a.** 192

    b. Answers will vary. One possibility is: stock shelves (140), mop (160), paint (200), scrub floors (300), mow grass (300)

Working Together

Answers will vary.

Unit 2

When Do I Add? p. 43

1–4. Answers will vary.

Talk About It

Answers will vary. Sample answers are 10 people, 6 vehicles, and 4 flowers.

Building an Addition Table pp. 44–45

Part A

+	0	1	2	3	4	5	6	7	8	9
0	0	1	2	3	4	5	6	7	8	9
1	1	2	3	4	5	6	7	8	9	10
2	2	3	4	5	6	7	8	9	10	11
3	3	4	5	6	7	8	9	10	11	12
4	4	5	6	7	8	9	10	11	12	13
5	5	6	7	8	9	10	11	12	13	14
6	6	7	8	9	10	11	12	13	14	15
7	7	8	9	10	11	12	13	14	15	16
8	8	9	10	11	12	13	14	15	16	17
9	9	10	11	12	13	14	15	16	17	18

Part B

1. $0 + 3 = 3$

   $3 + 0 = 3$

   $1 + 2 = 3$

   $2 + 1 = 3$

2. The answers are the same. For any addition fact, the numbers can be added in any order without changing the value of the sum.

Part C

3. $0 + 5 = 5$

   $1 + 4 = 5$

   $2 + 3 = 5$

4. $0 + 6 = 6$

   $1 + 5 = 6$

   $2 + 4 = 6$

   $3 + 3 = 6$

5. $0 + 7 = 7$

   $1 + 6 = 7$

   $2 + 5 = 7$

   $3 + 4 = 7$

6. $0 + 8 = 8$

   $1 + 7 = 8$

   $2 + 6 = 8$

   $3 + 5 = 8$

   $4 + 4 = 8$

7. $0 + 9 = 9$

   $1 + 8 = 9$

   $2 + 7 = 9$

   $3 + 6 = 9$

   $4 + 5 = 9$

8. $1 + 9 = 10$

   $2 + 8 = 10$

   $3 + 7 = 10$

   $4 + 6 = 10$

   $5 + 5 = 10$

Part D

9. $2 + 9 = 11$
$3 + 8 = 11$
$4 + 7 = 11$
$5 + 6 = 11$

10. $3 + 9 = 12$
$4 + 8 = 12$
$5 + 7 = 12$
$6 + 6 = 12$

11. $4 + 9 = 13$
$5 + 8 = 13$
$6 + 7 = 13$

12. $5 + 9 = 14$
$6 + 8 = 14$
$7 + 7 = 14$

13. $6 + 9 = 15$
$7 + 8 = 15$

14. $7 + 9 = 16$
$8 + 8 = 16$

15. $8 + 9 = 17$

16. $9 + 9 = 18$

Part E

Explanations will vary.

17. $5 + 7 = 12$
$5 + 7 = 12$
$5 + 7 = 12$

18. $2 + 6 = 8$
$2 + 6 = 8$
$2 + 6 = 8$

19. $3 + 0 = 3$
$3 + 0 = 3$
$3 + 0 = 3$

20. $9 + 1 = 10$
$9 + 1 = 10$
$9 + 1 = 10$

21. $1 + 1 = 2$
$2 + 2 = 4$
$3 + 3 = 6$
$4 + 4 = 8$
$5 + 5 = 10$
$6 + 6 = 12$
$7 + 7 = 14$
$8 + 8 = 16$
$9 + 9 = 18$

Addition Facts pp. 46–47

Part A

1. $3 + 6 = 9$
2. $4 + 4 = 8$
3. $3 + 4 = 7$
4. $2 + 5 = 7$
5. $2 + 7 = 9$
6. $9 + 0 = 9$

Part B

7. $8 + 6 = 14$
$4 + 3 = 7$
$6 + 3 = 9$
$1 + 1 = 2$
$5 + 9 = 14$
$4 + 9 = 13$
$3 + 5 = 8$
$2 + 5 = 7$
$8 + 7 = 15$
$3 + 7 = 10$
$8 + 0 = 8$

8. $6 + 7 = 13$
$5 + 8 = 13$
$1 + 5 = 6$
$4 + 4 = 8$
$2 + 8 = 10$
$0 + 2 = 2$
$7 + 3 = 10$
$2 + 4 = 6$
$7 + 1 = 8$
$8 + 4 = 12$
$7 + 4 = 11$

8. continued
$6 + 5 = 11$
$5 + 1 = 6$
$5 + 2 = 7$
$7 + 5 = 12$
$5 + 3 = 8$
$5 + 6 = 11$
$7 + 9 = 16$
$0 + 1 = 1$
$6 + 8 = 14$
$3 + 2 = 5$
$8 + 3 = 11$
$6 + 4 = 10$
$5 + 7 = 12$
$1 + 2 = 3$
$9 + 0 = 9$
$0 + 0 = 0$
$9 + 6 = 15$
$5 + 0 = 5$
$3 + 9 = 12$

9. $4 + 2 = 6$
$7 + 6 = 13$
$4 + 6 = 10$
$6 + 2 = 8$
$9 + 4 = 13$
$7 + 8 = 15$
$2 + 7 = 9$
$9 + 9 = 18$
$9 + 3 = 12$
$1 + 4 = 5$
$8 + 2 = 10$

10. $8 + 1 = 9$
$4 + 1 = 5$
$2 + 9 = 11$
$6 + 1 = 7$
$8 + 5 = 13$
$6 + 0 = 6$
$5 + 4 = 9$
$8 + 9 = 17$
$9 + 2 = 11$
$0 + 3 = 3$
$9 + 5 = 14$
$3 + 6 = 9$
$6 + 6 = 12$
$9 + 7 = 16$
$4 + 0 = 4$
$4 + 8 = 12$
$2 + 7 = 9$
$1 + 9 = 10$
$1 + 3 = 4$
$1 + 7 = 8$
$3 + 4 = 7$
$2 + 6 = 8$
$2 + 1 = 3$
$7 + 7 = 14$
$8 + 8 = 16$
$9 + 1 = 10$
$3 + 3 = 6$
$4 + 7 = 11$
$6 + 9 = 15$
$1 + 6 = 7$

Addition Strategies pp. 48–49

Part A

1. $10 + 6 = 16$
$8 + 10 = 18$
$10 + 1 = 11$
$10 + 2 = 12$

2. $7 + 9 = 16$
$9 + 3 = 12$
$5 + 9 = 14$
$8 + 9 = 17$

3. $3 + 4 + 6 = 13$
$8 + 7 + 2 = 17$
$5 + 9 + 5 = 19$

4. 15   11   14   10

Part B

5. $8 + 5 + 2 = \textbf{15}$

   $2 + 8 + 5 = \textbf{15}$

   $2 + 4 + 6 = \textbf{12}$

   $6 + 4 + 2 = \textbf{12}$

   $7 + 8 + 3 = \textbf{18}$

   $3 + 7 + 8 = \textbf{18}$

   $5 + 1 + 9 = \textbf{15}$

   $1 + 9 + 5 = \textbf{15}$

6. No. The commutative property of addition states that numbers can be added in any order without changing the value of the sum.

Part C

You can write the four equations in any order in problems 7 and 8.

7. **a.** $3 + 6 = 9$

   b. $6 + 3 = 9$

   c. $9 = 3 + 6$

   d. $9 = 6 + 3$

8. **a.** $7 + 8 = 15$

   b. $8 + 7 = 15$

   c. $15 = 7 + 8$

   d. $15 = 8 + 7$

Part D

9. $n = 8$

10. $A = 3$

11. $G = 6$

12. $t = 13$

13. $y = 5$

14. $x = 4$

15. $8 + 9 = f$

    $f = \textbf{17 fish}$

16. $\$5 + x = \$7$ per hour

    $x = \textbf{\$2}$

Adding Larger Numbers pp. 50–51

Part A

Estimates will vary. Exact answers are given below.

1. $46 $99 $956   787 $5,449

2. 79 $496   519 $1,488   999

Part B

3. $399   787 $1,678 $774   888

4. Discussion will vary. A sample answer is given below.

   You add the ones first in case there are more than 9 ones. If there are more than 9 ones, you can regroup to the tens place, and so on. It makes sense to add from right to left. (Regrouping is covered in the next lesson.)

Part C

5. **144 copies**

   $24 + 120 = c$

   $c = 144$

6. **268 miles**

   $134 + 134 = m$

   $m = 268$

7. **$385**

   $\$350 + \$35 = r$

   $r = \$385$

8. **683 videos**

   $103 + 220 + 310 + 50 = N$

   $N = 683$

Making Connections: Buying a Car p. 51

1. $25,300

2. Answers will vary, depending on whether the amounts are rounded to tens, hundreds, or thousands. A sample estimate is:

   $\$25,000 + \$1,000 + \$1,000 + \$100 + \$1,000 + \$300 = \$28,400$

3. $28,798

4. $29,798

Adding with Regrouping pp. 52–53

Part A

Estimates will vary. Exact answers are given below.

1. $83   871 $905 $3,292   1,172

2. 76 $914   8,331 $529

Part B

3. $662   5,697   1,032 $371

Part C

4. **156 presents**

   $46 + 18 + 39 + 53 = t$

   $t = 156$

5. **$4,114 per month**

   $\$1,750 + \$2,364 = s$

   $s = \$4,114$

6. **a. 1,027 pounds**

   $186 + 225 + 256 + 192 + 168 = w$

   $w = 1,027$

   b. 1,409 pounds

   $1,027 + 195 + 187 = w$

   $w = 1,409$

7. **1,525 people**

   $428 + 376 + 721 = p$

   $p = 1,525$

Making Connections: Maps and Distances p. 53

1. **548 miles; no**

   $197 + 94 + 257 = m$

   $m = 548$

2. **924 miles**

   $162 + 300 + 300 + 162 = m$

   $m = 924$

3. **1,208 miles**

   $197 + 94 + 257 + 198 + 300 + 162 = t$

   $t = 1,208$

Finding Perimeter pp. 54–55

Part A

1. 41 in.   8 cm

2. 344 in.   223 m

Part B

3. square   4. rectangle

   triangle   square

   rectangle   triangle

   polygon (or pentagon)   rectangle

Part C

5. 31 m   723 ft.   316 mi.

6. 348 mm   72 yd.   44 in.

Part D

7. **660 yards**

   $110 + 220 + 110 + 220 = 660$

8. **5,772 feet**

   $2,160 + 726 + 2,160 + 726 = 5,772$

9. **24 feet**

   $10 + 6 + 8 = 24$

10. **360 feet**

    $90 + 90 + 90 + 90 = 360$

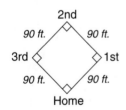

Mixed Review pp. 56–57

Part A

1. **a.** $2 + 3 = \mathbf{5}$ $4 + \mathbf{6} = 10$ $\mathbf{5} + 6 = 11$

   b. $7 + 4 = \mathbf{11}$ $\mathbf{8} + 5 = 13$ $7 + \mathbf{8} = 15$

   c. $8 + 0 = \mathbf{8}$ $9 + 9 = 18$ $6 + 3 = \mathbf{9}$

   d. $9 + 6 = \mathbf{15}$ $\mathbf{9} + 1 = 10$ $8 + 3 = \mathbf{11}$

2. Any of the following combinations are correct. Remember: the order of the numbers being added doesn't change the sum.

   $0 + 16 = 16$

   $1 + 15 = 16$

   $2 + 14 = 16$

   $3 + 13 = 16$

   $4 + 12 = 16$

   $5 + 11 = 16$

   $6 + 10 = 16$

   $7 + 9 = 16$

   $8 + 8 = 16$

3. **a.** 17   13   18   **b.** 16   19   16

4. 74   409   500   565   2,634

5. 87   565   136

Part B

6. $N = 14$ $C = 8$ $V = 15$ $Y = 4$

7. Any of the following equations are correct:

   $8 + 6 = 14$

   $6 + 8 = 14$

   $14 = 8 + 6$

   $14 = 6 + 8$

8. **96 exposures**

   $36 + 36 + 24 = 96$

Part C

9. **48 inches**

   $12 + 12 + 12 + 12 = 48$

10. **76 inches**

    $12 + 12 + 14 = 38$

    $38 + 38 = 76$

11. **70 inches**

    $15 + 15 + 20 + 20 = 70$

Part D

12. When you add a number to itself, the answer is an even number. Some examples are:

    $3 + 3 = 6; 4 + 4 = 8; 7 + 7 = 14;$

    $10 + 10 = 20$

13. **a.** When you add zero to a number, you get the same number.

   b. When 10 is added to a one-digit number, the 1 goes in the tens place and the other digit goes in the ones place.

   c. When 9 is added to a number, the total value is 1 less than 10 plus that value.

14. $1 + 9 = 10; 2 + 8 = 10; 3 + 7 = 10;$
    $4 + 6 = 10; 5 + 5 = 10$

15. Answers will vary.

Adding Thousands pp. 58–59

Part A

   1. 64,887   838,299   7,012,580

   2. 141,862   1,776,092   1,216,643

Part B

   3. 20,602 feet

   $20,320 + 282 = 20,602$

   4. $70,514

   $9,899 + $14,565 + $28,394 + $17,656 = $70,514

Part C

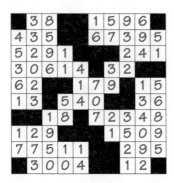

Understanding Word Problems pp. 60–61

Part A

   1. How many envelopes should he order . . . ?

   2. Find the total number of lunches served on Monday, Wednesday, and Friday. . . .

   3. How many cups of punch will a double recipe make?

   4. a. . . . how much money was pledged?

   b. Did the community reach its goal of $1,000 . . . ?

   5. Answers will vary.

Part B

Example 1: 33 hours

$6 + 6 + 6 + 8 + 7 = 33$

Example 2: 108 feet

$36 + 36 + 18 + 18 = 108$

Example 3: 567 miles

$186 + 283 + 98 = 567$

Example 4: There are many combinations possible. Two possible combinations are given here.

electricity:	$76
phone:	38
rent:	575
insurance:	+ 108
	$797

rent:	$575
food and entertainment:	+ 225
	$800

Part C

   1. 750 envelopes

   $375 + 289 = 664$

   Although the total needed is 664, the smallest box of envelopes needed is 750.

   2. 144 lunches

   43 (Monday) $+ 39$ (Wednesday) $+ 62$ (Friday) $= 144$

   3. 28 cups

   $4 + 2 + 8 = 14$

   $14 + 14 = 28$

   4. a. $736

   $56 + $185 + $350 + $145 = $736

   b. No

   $736 < $1,000

   5. Answers will vary.

Adding on a Calculator pp. 62–63

Part A

   1. 0. *or* 0

   2. Answers will vary.

   3. 0. *or* 0

   4. a. 56

   b. 57

   c. 80; 56 was erased and 23 and 57 were added.

Part B

Estimates will vary. Any reasonable estimate is correct.

5. Estimate: 90 + 250 = 340

   Answer: 332

6. Estimate: 2,500 + 7,000 = 9,500

   Answer: 9,440

7. Estimate: 50,000 + 50,000 = 100,000

   Answer: 101,762

Part C

8. 2,568

9. 340

10. 6,127,116

Adding Dollars and Cents pp. 64–65

Part A

1. $60 $335   91¢

2. $3,271 $1.42 $2.20

3. $17.67 $500.45 $19.96

Part B

4. $84.24 $419.15 $1.20

5. 60¢ $137.45 $33

Part C

6. $28,300

7. $517.32 (The 8% tax rate in the problem is not needed to solve this question.)

8. $45.30

9. $1,104.83

Making Connections: Ordering from a Catalog p. 65

1.

Qty	Item Number	Description	Price Each	Total
2	34207	mini-chopper	$19.95	$39.90
1	14752	wok	25.99	25.99
2	30521	wooden utensils	3.49	6.98

Kitchen & Cooks

Merchandise Total	$72.87
Shipping & Handling	3.95
Total	$76.82

2. **Camping Warehouse**

Qty	Item Number	Description	Price Each	Total
1	42111	tent	$298.99	$298.99
1	21475	sleeping bag	58.25	58.25
1	74281	lantern	19.99	19.99

Merchandise Total	$377.23
Shipping & Handling	7.50
Total	$384.73

Estimating Costs pp. 66–67

Part A

1. a. $10.80   2. a. $1,080

   b. $21.60   b. $4,320

Part B

Estimates will vary. Any reasonable estimate is correct.

3.

	Exact	Estimate
1 whole chicken	$3.98	$4.00
5 pounds potatoes	1.49	1.50
1 head lettuce	.98	1.00
1 tray tomatoes	1.29	1.00
1 pound frozen peas	.89	1.00
6-pack cola	1.69	2.00
1 gallon milk	2.59	2.50
1 pound wheat bread	1.29	1.00
Subtotal	$14.20	$14.00
Tax	1.14	1.00
Total	$15.34	$15.00

4.

	Exact	Estimate
smoke alarm	$3.99	$5.00
cordless drill	39.00	40.00
1 pound screws	4.78	5.00
table saw	165.99	170.00
Subtotal	$213.76	$220.00
Tax	17.10	20.00
Total	$230.86	$240.00

5.

	Exact	Estimate
watch	$44.99	$45.00
wallet	9.88	10.00
gold chain	149.99	150.00
bread maker	179.99	200.00
cordless phone	78.88	80.00
Subtotal	$463.73	$485.00
Tax	37.10	40.00
Total	$500.83	$525.00

Unit 2 Review pp. 68–69

Part A

1. a. 15   b. 20   c. 3

2. c, a, b, d

3. 182 < 376   4,506 > 4,056   8 + 7 = 7 + 8

4. c, a, d, b

5. 4:15   10:30   1:40

6. a. 63   **b.** 75   **c.** 88

7. a. 300   **b.** $70

Part B

8. 16   39   120   458   10,121

9. 16   51   1,259

10. 1,166,368   97,662 $84.21 $45.36

11. a. $N = 7$ **b.** $t = 9$ **c.** $y = 41$

Part C

12. 72 feet

Question: How many feet of molding does she need?

$18 + 18 + 18 + 18 = 72$

13. Yes, she can afford to buy the items. The total is $62.53, which is less than $65.00.

Question: Can she afford to buy [the items listed]?

$12.95 + $19.95 + $25.00 + $4.63 = $62.53

14. subtotal: **$53.93**

total: **$58.20**

Working Together

Answers will vary.

Unit 3

When Do I Subtract? p. 71

Answers will vary.

Subtraction Strategies pp. 72–73

Part A

1. $4 - 1 = 3$ $6 - 4 = 2$ $10 - 3 = 7$

   $4 - 3 = 1$ $6 - 2 = 4$ $10 - 7 = 3$

2. $7 - 4 = 3$ $8 - 6 = 2$ $9 - 3 = 6$

   $7 - 3 = 4$ $8 - 2 = 6$ $9 - 6 = 3$

3. $11 - 4 = 7$ $14 - 6 = 8$ $12 - 7 = 5$

   $11 - 7 = 4$ $14 - 8 = 6$ $12 - 5 = 7$

Part B

4. a. $14 - 5 = 9$

   b. $16 - 8 = 8$

   c. $11 - 4 = 7$

Part C

5. a. 9   5   0   **b.** 3   3   0

6. $8 - 8 = \mathbf{0}$ $7 - \mathbf{1} = 6$ $5 - \mathbf{0} = 5$

   $8 - \mathbf{8} = 0$ $7 - 1 = \mathbf{6}$ $\mathbf{5} - 0 = 5$

   $\mathbf{8} - 8 = 0$ $\mathbf{7} - 1 = 6$ $5 - 5 = \mathbf{0}$

7. $5 + 8 = 13$

   $8 + 5 = 13$

   $13 - 5 = 8$

   $13 - 8 = 5$

Subtraction Facts pp. 74–75

Part A

1. 5	0	5	0
2. 5	1	0	3
3. 5	1	4	6
4. 7	1	4	6
5. 0	0	1	2
6. 3	7	0	0
7. 0	0	3	6
8. 1	1	9	2
9. 1	4	8	2
10. 1	2	4	2
11. 3	2	4	0
12. 3	1	5	5
13. 7	3	2	4
14. 2	6	3	8

Part B

15. 9	3	4	9
16. 9	9	4	5
17. 8	9	9	8
18. 6	7	7	7
19. 2	6	8	8
20. 9	6	3	7

Part C

21. $12 - 7 = 5$ **24.** $8 - 2 = 6$

22. $9 - 5 = 4$ **25.** $10 - 7 = 3$

23. $11 - 4 = 7$

Part D

26. $r = 7$ $Q = 13$ b is any number.

27. $N = 15$ $m = 0$ $c = 9$

Part E

28. True   **33.** True

29. False   **34.** True

30. True   **35.** False

31. True   **36.** True

32. False   **37.** False

Subtracting Larger Numbers pp. 76–77

Part A

Estimates will vary.

1. 26	22	136	211	7,110
2. 80	32	511	251	
3. 523	601	342	424	

Part B

Answers will vary. Sample answers are given below.

4. A placeholder is a zero that keeps a place value equal to zero. Placeholder zeros are necessary to keep the value of the answer correct.

5. The zero in 302 means there are no tens in the answer. The zero in 130 means there are no ones in the answer. So they don't mean the same thing. One zero holds the tens place empty and the other zero holds the ones place empty.

Part C

6. 330	21	2	7,108	3,200
7. 622	6	110	4,000	10

Part D

Estimates will vary.

8. **1971**

$1994 - 23 = 1971$

9. **$1,574**

$\$1,997 - \$423 = \$1,574$

10. **$40,740**

$\$47,995 - \$7,255 = \$40,740$

11. **13 miles**

$578 - 565 = 13$

Explanations will vary. You are finding a difference.

Deciding to Add or Subtract pp. 78–79

Estimates will vary.

1. a. How much more will he have to pay?

b. You can use subtraction because you're finding the difference between the down payment and the total cost of the car. (You could also think in terms of addition. $2,525 plus what amount equals $13,869?)

c. $\$14,000 - \$3,000 = \$11,000$

or $\$14,000 - n = \$3,000$

$n = \$11,000$

d. $11,344; yes, it's sensible

2. a. How many yards has Denver gained?

b. Use subtraction because Denver lost yards after their original gain of 9 yards.

c. 5 yards

d. 4 yards; yes, it's sensible

3. a. What is the new bank balance?

b. Use addition because you're combining money amounts.

c. $\$200 + \$100 = \$300$

d. $313; yes, it's sensible

Making Connections: Planning a Vacation p. 79

1. $156

$\$78 + \$42 + \$36 = \$156$

2. 411 miles

$152 + 287 = 439$

$439 - 28 = 411$

3. $78

$\$24 + \$24 + \$15 + \$15 = \$78$

4. 144 miles

$564 - 420 = 144$

5. 183 Texans

$2 + 2 = 4$

$187 - 4 = 183$

6. Answers will vary.

7. 732 miles; 532 miles shorter

$150 + 209 + 158 + 215 = 732$

$732 - 200 = 532$

8. 2,428 miles

$28 + 152 + 287 + 420 + 564 + 150 + 209 + 158 + 215 + 245 = 2,428$

9. 1,151 fewer miles

$3,579 - 2,428 = 1,151$

Subtracting by Regrouping pp. 80–81

Part A

Estimates will vary.

1. 18	8	44	46	19
2. 271	592	92	6,911	5,621

Part B

3. 55	909	62	38	191
4. 7,003	7,900	900	690	

Part C

Answers will vary.

5. $507 + 79 = 586$

$79 + 507 = 586$

$586 - 507 = 79$

$590 - 80 = 510$ (estimate)

6. 2,791

7. $1,857 - 283 = 1,574$

$1,857 - 1,574 = 283$

8. a. false

b. true

c. true

d. false

9. Estimate: $300 - 100 = 200$

Exact: 192

Mixed Review pp. 82–83

Part A

1.	340	701	9,131	4,251
2.	9,013	571	130	
3.	49	707	109	5,901
4.	591	6,533	192	3,110
5.	863	4,092	3,911	

Part B

6.	510	1,921	60,421	71,920
7.	80	56,201	699	
8.	31,090	139	60,601	

Part C

9. $65 - 28 = 37$

$65 - 37 = 28$

10. $219 + 75 = 294$ or $75 + 219 = 294$

11. Examples will vary.

a. The number stays the same. Example: $9 - 0 = 9$

b. The answer is 1 less than the number. Example: $9 - 1 = 8$

c. The answer is 0. Example: $9 - 9 = 0$

12. a. $48 - 12 = 36$

b. $60 - 19 = 41$

13. a. $N = 22$

b. $r = 27$

c. $A = 92$

Part D

14. **$2,591**

$\$2,346 + \$245 = \$2,591$

15. **$3,911**

$\$7,258 - \$3,347 = \$3,911$

16.

Hometown Savings Bank			
Date	Deposit	Withdrawal	Balance
1/1	$1,200		$1,200
3/1	$250		1,450
5/1	$145		1,595
5/20		$143	1,452
7/1	$378		1,830
8/1		$1,595	235
8/20	$157		392
10/1		$108	284
11/1	$288		572
12/30		$72	500

Regrouping More than Once pp. 84–85

Part A

1.	288	87	5,867	8,798
2.	289	3,809	59,069	

Part B

3.	8,667	88,490	2,903	6,075

Part C

4.

Item	Original	Sale	Savings
19" Color TV with Remote	$349	$299	$50
13" Color TV	$189	$129	$60
AM/FM Cassette Player	$195	$158	$37
Projection TV	$1,788	$1,499	$289
VCR	$229	$189	$40

Making Connections: Orbiting the Sun p. 85

1. Mercury

2. **2,771 million miles**

$3,658$ million $- 887$ million $= 2,771$ million

3. a. colder; because it's farther from the Sun

b. **1,687 million miles**

$1,780$ million $- 93$ million $= 1,687$ million

4. **60,000 days**

$60,225 - 225 = 60,000$

5. **90,155 days**

$90,520 - 365 = 90,155$

6. It takes so long for Pluto to orbit the Sun because it travels a great distance.

What Do I Need to Find? pp. 86–87

Part A

1. . . . how many more miles did she have to travel to get to Appleton?

   To solve, find the difference between the distance in miles to Appleton and Fond du Lac by subtracting $101 - 62 = 39$.

2. **a.** What was the total number of students in 1993?

   To solve, add the numbers of men and women students in 1993: $12,499 + 12,031 = 24,530$

   b. How many women attended Harper in 1994?

   To solve, find the difference between total number of students and the number of men students in 1994: $24,572 - 11,943 = 12,629$

3. **a.** How much weight has he lost so far?

   To solve, add the number of pounds lost: $5 + 3 + 3 + 0 + 1 = 12$

   b. . . . how much weight does he have left to lose?

   To solve, find the difference between weight he wants to lose and weight lost already: $35 - 12$ (pounds already lost) $= 23$

Part B

4. **$2,730**

   $\$1,800 + \$930 = \$2,730$

5. **$870**

   $\$1,800 - \$930 = \$870$

6. **$1,655**

   $\$1,800 - \$145 = \$1,655$

7. **33 lures**

   You could have solved the problem in either of these ways:

   $130 - 97 = 33$

   or

   $97 + x = 130$

   $x = 33$

Subtracting from Zeros pp. 88–89

Part A

1. 464   3,119   857   5,404
2. 1,193   6,125   1,989   428

Part B

3. 55   5,409   13,360

Part C

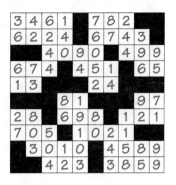

Subtracting Dollars and Cents pp. 90–91

Part A

1. $54 $202   8¢
2. $5,858 $3.17 $14.67
3. $.03 $41.24 $1.16

Part B

4. $59 $18.75 $.09
5. $8.01 $.37 $3.20
6. $67,950 $7.12 $725

Part C

7. **$98**

   $\$248 - \$150 = \$98$

8. **$9.60**

   $\$10.00 - \$0.40 = \$9.60$

9. **$184.25**

   $\$1,150 - \$965.75 = \$184.25$

10. **$200.63**

    $\$275.63 - \$75 = \$200.63$

Making Connections: Tracking Vacation Spending p. 91

Date	Description	Expenses	Balance
8/5	Beginning balance		$1,500.00
8/5	Gas for car	$15.40	$1,484.60
8/5-8/12	Lodging	$450.00	$1,034.60
8/5	Fishing license	$27.50	$1,007.10
8/6	Bait	$9.65	$997.45
8/6-8/12	Boat rental	$182.00	$815.45
8/6-8/12	Boat gasoline	$23.72	$791.73
8/12	Souvenirs	$86.35	$705.38
8/5-8/12	Food	$280.00	$425.38
8/12	Gas for car	$14.88	$410.50
8/13	Ending balance		$410.50

Understanding Your Paycheck pp. 92–93

Part A

1. $756.50

2. $529.77

3. **$226.73**

   $756.50 − $529.77 = $226.73

4. **$4.25**

   $12.75 − $8.50 = $4.25

5. $756.50 − $226.73 = $529.77

6. 86

7. $680.00 + $76.50 = $756.50

8. **$211.83**

   $12.40 + $2.50 = $14.90

   $226.73 − $14.90 = $211.83

Part B

CUSTER CONSTRUCTION COMPANY		NO. 116735	
EMPLOYEE NAME: Joanne May SOCIAL SECURITY #: 330-46-2706		EXEMPTIONS: Fed 1 State 1 Status 1 PAY PERIOD: 1/29 to 2/11	
CURRENT EARNINGS		DEDUCTIONS: Current/Year-to-Date	
	Hours/Rate/Pay	FEDERAL	143.06 401.63
REGULAR	80 8.50 680.00	STATE	15.90 44.46
OVERTIME	9 12.75 114.75	FICA	55.63 156.19
PAY	Current/Year-to-Date	INSURANCE	15.90 30.27
GROSS	794.75 2,231.25	UNION	12.40 37.20
NET	549.36 1,554.00	UNITED WAY	2.50 7.50
VACATION BALANCE: 10 DAYS		TOTAL	245.39 677.25

Subtracting on a Calculator pp. 94–95

Part A

1. 93   777   46,846   151

2. $4.02 $4.64 $41.35 $627.20

Part B

Estimates will vary.

3. 524 − 67 = n

   Estimate: n ≈ 400

   Exact: n = 457

4. 3,472 − 2,905 = W

   Estimate: W ≈ 500

   Exact: W = 567

5. 604 − 575 = d

   Estimate: d ≈ 25

   Exact: d = 29

6. $733.60 − $485.50 = L

   Estimate: L ≈ $200

   Exact: L = $248.10

Part C

7. $5.32

Figuring Change pp. 96–97

Part A

1. $5.28 $11.05 $2.55

2. $17.21 $13.50 $8.94

Part B

Note: If there are no cents to be entered in a calculator, you do not need to enter the zeros.

3. **$1.75**

   $5.00 − $3.25 = $1.75

4. **$35**

   $122.25 − $87.25 = $35.00

5. 32¢ = **$0.32**

   68¢ = $0.68

   $1.00 − $0.68 = $0.32

6. **$5.50**

   $40.00 − $34.50 = $5.50

Part C

Coins and bills may vary. Sample answers are given below.

7. a. $14.76   b. $10, $1, $1, $1, $1, 25¢, 25¢, 25¢, 1¢

8. a. 89¢ = $.89   b. 25¢, 25¢, 25¢, 10¢, 1¢, 1¢, 1¢, 1¢

9. a. $24.41   b. $10, $10, $1, $1, $1, $1, 25¢, 10¢, 5¢, 1¢

10. a. $18.45   b. $10, $5, $1, $1, $1, 25¢, 10¢, 10¢

Part D

Answers will vary.

Unit 3 Review pp. 98–99

Part A

1. 7

2. sample answer: 456

3. 0, 2, 4, 6, 8 or any number with one of these as a last digit

4. 7

5. 3

6. any number 0 through 16

7. 12

8. any number greater than or equal to 4,358

9. $400

10. 365

11. 9:30 A.M.

12. 0

13. 4

14. 3

15. $A = 9$

16. 52

17. 462

18. 16,337

19. 42 feet

20. $1.35

Part B

21. a. 17 0 5 8   **b.** 5 11 8 8

22. 121 60 1,004 587 6,489

23. 37 586 709

24. 553 1,129 18,099

25. $t = 21$

$n = 11$

$D = 15$

Part C

26. 1969

1994 − 25 = 1969

27. $3,089

$2,846 + $350 + $189 = $3,385

$3,385 − $296 = $3,089

28. $11,850

$12,600 − $750 = $11,850

29. $12.62

$20 + $20 + $20 = $60

$60 − $47.38 = $12.62

Answers on bills and coins will vary.

$10, $1, $1, 25¢, 25¢, 10¢, 1¢, 1¢

30. $1,625

$15,500 − $13,875 = $1,625

Working Together

Answers will vary.

Unit 4

When Do I Multiply? p. 101

1. A

2. M

3. M

4. S

5. M

6. **a.** 3 × 4   **d.** $56 × 2

   b. not possible   **e.** 6 × 4

   c. not possible

7. 2 4 6 8 10 **12 14 16** 18 20

   5 10 15 **20 25** 30 **35 40 45** 50

Talk About It

Answers will vary.

Building a Multiplication Table pp. 102–103

Part A

×	0	1	2	3	4	5	6	7	8	9	10
0	0	0	0	0	0	0	0	0	0	0	0
1	0	1	2	3	4	5	6	7	8	9	10
2	0	2	4	6	8	10	12	14	16	18	20
3	0	3	6	9	12	15	18	21	24	27	30
4	0	4	8	12	16	20	24	28	32	36	40
5	0	5	10	15	20	25	30	35	40	45	50
6	0	6	12	18	24	30	36	42	48	54	60
7	0	7	14	21	28	35	42	49	56	63	70
8	0	8	16	24	32	40	48	56	64	72	80
9	0	9	18	27	36	45	54	63	72	81	90
10	0	10	20	30	40	50	60	70	80	90	100

Part B

1. **3 × 8** = 24

   8 × 3 = 24

   4 × 6 = 24

   6 × 4 = 24

2. 3 × 0 = **0**

   7 × **0** = 0

   any number × 0 = 0

   2 × **0** = 0

3. 6 × 1 = **6**

   9 × 1 = 9

   4 × 1 = **4**

   5 × 1 = 5

   8 × **1** = 8

4. 2 × 0 = **0**

   2 × 1 = **2**

   2 × 2 = 4

   2 × 3 = **6**

   2 × 4 = **8**

   2 × 5 = 10

   2 × 6 = 12

   2 × 7 = **14**

   2 × 8 = 16

   2 × **9** = 18

   2 × 10 = **20**

5. 3 × **0** = 0

   3 × 1 = 3

   3 × 2 = **6**

   3 × **3** = 9

   3 × 4 = 12

   3 × **5** = **15**

   3 × 6 = 18

   3 × 7 = **21**

   3 × 8 = 24

   3 × **9** = 27

   3 × 10 = **30**

6. 4 × **1** = 4

   4 × 2 = **8**

   4 × 3 = **12**

   4 × 4 = 16

   4 × 5 = 20

   4 × 6 = **24**

   4 × **7** = 28

   4 × 8 = **32**

   4 × 9 = **36**

   4 × **10** = 40

7. $5 \times \mathbf{0} = 0$

$5 \times 1 = \mathbf{5}$

$5 \times 2 = 10$

$5 \times \mathbf{3} = 15$

$5 \times 4 = \mathbf{20}$

$5 \times \mathbf{5} = 25$

$5 \times \mathbf{6} = 30$

$5 \times \mathbf{7} = 35$

$5 \times \mathbf{8} = 40$

$5 \times 9 = \mathbf{45}$

$5 \times \mathbf{10} = \mathbf{50}$

8. $6 \times \mathbf{1} = 6$

$6 \times 2 = \mathbf{12}$

$6 \times 3 = \mathbf{18}$

$6 \times \mathbf{4} = 24$

$6 \times 5 = \mathbf{30}$

$6 \times \mathbf{6} = 36$

$6 \times 7 = \mathbf{42}$

$6 \times 8 = \mathbf{48}$

$6 \times 9 = 54$

$6 \times \mathbf{10} = 60$

Part C

9. a. Multiples of 5 end in 5 or 0.

b. Multiples of 10 end in 0.

10. The digits add to nine.

Consecutive multiples begin with consecutive digits.

Each answer begins with a digit that is one less than the number being multiplied by 9.

11. $8 \times 6 = \mathbf{48}$ $8 \times 7 = \mathbf{56}$ $9 \times 8 = \mathbf{72}$

$9 \times 7 = \mathbf{63}$ $7 \times 6 = \mathbf{42}$ $9 \times 6 = \mathbf{54}$

Multiplication Facts pp. 106–107

Part A

1. a. 12	**b.** 36	
32	12	
63	18	
2. a. 0	**b.** 25	
24	0	
14	20	
3. a. 24	**b.** 42	
64	40	
54	8	
4. a. 36	**b.** 48	
15	16	
28	45	
5. a. 30	**b.** 35	
10	18	
72	63	
6. a. 63	**b.** 1	
16	6	
48	27	
7. a. 10	**b.** 30	
32	0	
15	49	
8. a. 40	**b.** 54	
35	18	
24	9	
9. a. 36	**b.** 28	
20	42	
8	48	
10. a. 45	**b.** 0	
12	32	
10	56	

Part C

9. $7 \times \mathbf{1} = 7$

$7 \times 2 = \mathbf{14}$

$\mathbf{7} \times 3 = 21$

$7 \times \mathbf{4} = \mathbf{28}$

$7 \times 5 = \mathbf{35}$

$7 \times \mathbf{6} = 42$

$7 \times 7 = \mathbf{49}$

$7 \times \mathbf{8} = \mathbf{56}$

$7 \times 9 = 63$

10. $8 \times \mathbf{2} = 16$

$8 \times 3 = \mathbf{24}$

$8 \times 4 = \mathbf{32}$

$8 \times \mathbf{5} = 40$

$\mathbf{8} \times \mathbf{6} = 48$

$8 \times 7 = \mathbf{56}$

$8 \times 8 = \mathbf{64}$

$8 \times 9 = 72$

$\mathbf{8} \times 10 = 80$

11. $9 \times \mathbf{2} = 18$

$9 \times 3 = \mathbf{27}$

$9 \times \mathbf{4} = 36$

$9 \times 5 = 45$

$9 \times 6 = \mathbf{54}$

$9 \times 7 = \mathbf{63}$

$9 \times 8 = 72$

$\mathbf{9} \times \mathbf{9} = 81$

$9 \times \mathbf{10} = 90$

Multiplication Strategies pp. 104–105

Part A

1. $5 \times \mathbf{2} = 10$

2. $3 \times 4 = 12$

3. $6 \times 1 = 6$

4. $2 \times 5 = 10$

5. $4 \times 3 = 12$

6. $1 \times \mathbf{6} = 6$

Part B

7. $3 \times 6 = \mathbf{6} \times 3$ $4 \times 2 = \mathbf{8}$ $6 \times 5 = \mathbf{30}$

$5 \times 2 = 2 \times \mathbf{5}$ $7 \times 2 = \mathbf{14}$ $5 \times 4 = \mathbf{20}$

8. $7 \times 6 = \mathbf{42}$ $9 \times 6 = \mathbf{54}$ $6 \times 10 = \mathbf{60}$

$6 \times 7 = 42$ $6 \times 9 = 54$ $\mathbf{10} \times 8 = 80$

11. a. 72
    4
    7

b. 8
    12
    81

12. a. 0
    2
    56

b. 27
    9
    18

Part B

13. $8 \times 5 = 40$ *or* $8(5) = 40$
14. $3 \times 8 = 24$ *or* $3(8) = 24$
15. $7 \times 7 = 49$ *or* $7(7) = 49$
16. $6 \times 8 = 48$ *or* $6(8) = 48$

Part C

17. $T = 28$
    $n = 10$
    $W = 56$

18. $y = 8$
    $c = 6$
    $G = 27$

Part D

19. $8n = 56$
20. $12 - 7 = M$
21. $6(7) = P$
22. $Q + 8 = 13$

Multiplying by One-Digit Numbers pp. 108–109

Part A

1. a. 369   48   66   **b.** 80   770   804
2. a. 155   1,206   75   **b.** 560   0   276

Part B

3. 2,406   276   729   5,600
4. 840   48,666   49,000   2,550

Part C

5. 240   99   88   0
6. 48   48   200   84

Part D

7. $37 per person
    $41 \times 7 = \$287$
    $\$287 - \$250 = \$37$

8.

×	20	91	501
7	140	637	3,507
3	60	273	1,503
6	120	546	3,006

9. 87 years
    $20 \times 4 = 80$
    $80 + 7 = 87$

10. $210
    $\$73 \approx \$70$
    $\$70 \times 3 = \$210$

Multiplying and Regrouping pp. 110–111

Part A

1. a. 258   315   94   **b.** 96   472   511
2. a. 3,008   3,150   4,221
    b. 1,672   2,349   850

Part B

3. a. 285   2,169   1,848
    b. 1,656   273   5,664
4. a. 2,442   9,129   48,560
    b. 36,081   35,567   7,042

Part C

5. 792   32,600   4,359   14,693   3,170
6. 22,644   65,424   3,500   72,150   6,216

Part D

7. Multiply by 3.

Input	8	10	17	48	175
Output	24	30	51	144	525

8. Multiply by 6.

Input	4	9	18	50	376
Output	24	54	108	300	2,256

Multiples of 10 pp. 112–113

Part A

1. 70   50   900   700   3,000
2. 450   120   51,100   48,100   94,000

Part B

3. a. 150   4,900   5,400
    b. 56,000   480   100
4. a. 1,800   8,100   20,000
    b. 240,000   3,600   3,500,000
5. 6,930   108,600   28,800   72,000   9,000

Part C

6. Estimate: $90 \times 50 = 4,500$; Exact answer: **(3)** 4,361
7. Estimate: $900 \times 60 = 54,000$;
    Exact answer: **(2)** 55,366
8. Estimate: $300 \times 500 = 150,000$;
    Exact answer: **(3)** 159,390

Part D

Estimates will vary.

9. $360

$88 \approx $90

$90 \times 4 = $360

10. 4,200

72 $\approx$ 70

70 $\times$ 60 = 4,200

This estimate is lower because you rounded down.

70 $\times$ 60 < 72 $\times$ 60

11. $30,000

$625 \approx $600

52 $\approx$ 50

$600 \times 50 = $30,000

12. $120,000

3,899 $\approx$ 4,000

$28 \approx $30

$30 \times 4,000 = $120,000

This estimate is higher because you rounded up.
$30 \times 4,000 > $28 \times 3,899

Multiplying by Two-Digit Numbers pp. 114–115

Part A

Estimates will vary. Exact answers are given below.

1. 23,033   1,400   48,672   13,224   54,384
2. 14,256   24,804   145,628   6,336

Part B

3. 30,738   11,880   45,760   311,535   4,935,315
4. 15,543   58,112   625   13,986

Part C

5. 432 cans

18 $\times$ 24 = 432

6. $1,300

$25 \times 52 = $1,300

7. 1,728 seats

36 $\times$ 48 = 1,728

8.

$\times$	35	42	18
18	630	756	324
35	1,225	1,470	630
42	1,470	1,764	756

Part D

9. $5,100

$425 \times 12 = $5,100

10. No, $225 a month for 36 months comes to only $8,100.

11. $21,600. Answers will vary. Look for different combinations.

What Information Do I Need? pp. 116–117

Part A

1. $38, twice (2)
2. 16 students, 50 words
3. 5 people, $4 per person, $25
4. 4 dozen (48), 40 boxes

Part B

1. $76

$38 \times 2 = $76

2. 800 words

16 $\times$ 50 = 800

3. yes

5 $\times$ $4 = $20 $20 < $25

4. They estimated too few boxes.

four dozen = 4 $\times$ 12 = 48

40 < 48

Part C

5. 27 $\times$ 40 = **1,080 calls**

6. $75 \times 14 = **$1,050**

7. 1,682 $\times$ $96 = **$161,472**

8. 240 $\times$ 5 $\times$ 2 = **2,400 cases**

Making Connections: Buying in Quantity p. 117

1. Answers will vary.
2. a. $1,160
   b. $1,050
3. Answers will vary. A sample answer is given below.

   Since the price per guest is less for 30 guests than 29 guests, it's cheaper to buy more tickets.

Mixed Review pp. 118–119

Part A

1. a. 72   42   72   b. 36   63   56
2. a. 81   54   48   b. 54   0   7
3. 16   56   24   28

4. 72   42   24   0

5. 90   3,000   600   8,000

6. a. 128   3,577   5,409
   b. 5,600   963   3,608

7. 48   210   0

Part B

8. a. 765   222   316
   b. 5,112   48,056   8,181

9. 3,200 27,000 9,700 303,500 2,760,000

10. 1,430 51,642 18,240 18,048 370,062

11. 16,940 9,398 17,200

12. 2,754 36,000 28,026

Part C

13. 9,100 miles

$175 \times 52 = 9,100$

14. $750

$45 - 40 = 5$

$\$15 \times 2 = \$30$

$\$15 \times 40 = \$600$

$\$30 \times 5 = \$150$

$\$600 + \$150 = \$750$

15. $k = 9$ $H = 56$ $t = 9$

16. $3,626

$98 \times \$37 = \$3,626$

Part D

17. 4,500 jackets

$1,500 \times 3 = 4,500$

18. $272,000

$4,000 \times \$68 = \$272,000$

19. a. November

   b. $170,000

20. Answers will vary. Answers could include sales for February, March, April, May, June, July, October, and December.

Multiplying by Larger Numbers pp. 120–121

Part A

1. 65,148 133,000 1,075,864 3,248,865

2. 266,178 681,750 1,531,635 18,392,505

3. 6,863,130 spectators

$42,365 \times 162 = 6,863,130$

Part B

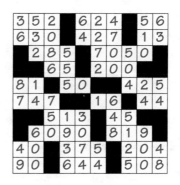

Area and Volume pp. 122–123

Part A

1. $12 \times 1 = 3 \times 4 = 2 \times 6 = 12$ square units

2.

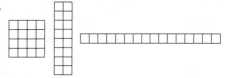

3. a. 49 square inches   **b.** 40 square miles
   c. 1,200 square yards

4. 6,600 square yards

Part B

5. 162 cubic inches   **6. a. 2,400 cubic feet**

$9 \times 3 \times 6 = 162$ $20 \times 15 \times 8 = 2,400$

                          b. 16,800 gallons

                          $7 \times 2,400 = 16,800$

7. yes

8. Answers will vary.

Multiplying Dollars and Cents pp. 124–125

Part A

1. $252 $10,000 $43,600

2. $47.70 $40,830 $10,534.08

3. 70¢ $21.36 $24

Part B

4. $392 $11,130 $391,950

5. $46.35 $2,146.56 $15,831.20

6. $34.20 $9 $18

Part C

7. $2.25

$75¢ = \$.75$

$3 \times .75 = 2.25$

8. $6,680.40

2 years = 24 months

$278.35 × 24 = $6,680.40

9. $1,249.50

$24.99 × 50 = $1,249.50

10. $1,184

$18.50 × 64 = $1,184

Making Connections: Taking Inventory p. 125

Eagle Sport Shop Inventory: January

Item	Tally	Total Items	Value per Item	Total Amount
Headgear	ЖЖ ЖЖ IIII	14	$11.45	$160.30
Sport socks	ЖЖ ЖЖ ЖЖ ЖЖ III	23	3.68	84.64
Jump ropes	ЖЖ ЖЖ ЖЖ III	18	1.95	35.10
Sport gum	ЖЖ ЖЖ ЖЖ ЖЖ ЖЖ ЖЖ	30	.37	11.10
Sweat suits	ЖЖ ЖЖ ЖЖ ЖЖ II	22	29.15	641.30
Treadmill	ЖЖ ЖЖ IIII	14	138.60	1,940.40

Total Value: $2,872.84

Multiplying on a Calculator pp. 126–127

Part A

1. 188,100   900,426   3,806,716
2. 44,148   386,048   656,668
3. 68,445   244,944   20,832
4. $1,232.10 $122.50 $23,005.44

Part B

Estimates will vary. Sample estimates are given below.

5. Estimate = 72,000   Estimate = 2,100,000

    Actual = 72,306   Actual = 2,464,760

6. Estimate = 36,000   Estimate = 75,000

    Actual = 40,824   Actual = 66,960

Part C

Item	Amount	Yearly Total
Rent	$386 per month	$4,632
Food	$75 per week	3,900
Car insurance	$1,064 per year	1,064
Health insurance	$28 per week	1,456
Utilities	$150 per month	1,800
Car payments	$247 per month	2,964
Gas	$96 per month	1,152
Phone	$35 per month	420
Miscellaneous	$100 per week	5,200

Total: $22,588

Multistep Problems pp. 128–129

Part A

1. **Step 1:** $12.45 × 40 = $498

   Step 2: $540 + $498 = $1,038

2. **Step 1:** $12.45 + $1.86 = $14.31

   Step 2: $14.31 × 40 = $572.40, which is $32.40 more than Gloria earns ($540).

Part B

3. **1,000 square feet**

   25 × 25 = 625

   15 × 25 = 375

   625 + 375 = 1,000

4. **$1,296**

   18 × 24 = 432

   432 × $3 = $1,296

5. **yes;** 20 gallons will cover 300 × 20 = 6,000 square feet. The warehouse floor is only 75 × 75 = 5,625 square feet.

Part C

Answers will vary.

Filling Out an Order Form pp. 130–131

Part A

Country Home Shop				
Catalog Number	Item Description	Quantity	Price Each	Price Total
A770-3366	Table lamps	2	49.99	$99.98
A745-4051	Woven rug	1	24.60	24.60
A736-8707	Mini-blinds	4	18.40	73.60
A727-3311	Pillow	2	19.99	39.98
			Merchandise Total:	$238.16
			Sales Tax:	16.67
			Delivery/Shipping Charges:	3.95
			Total:	$258.78

Part B

1. **$63.95**

   $7.99 × 3 = $23.97

   $19.99 × 2 = $39.98

   $23.97 + $39.98 = $63.95

2. **$17.55**

   $10.36 + $7.19 = $17.55

3. $165.49

$12.00 + $23.97 + $39.98 + $39.99 + $32.00 = $147.94

$147.94 + $17.55 = $165.49

4. $107.95

new merchandise total = old merchandise total – jacket price = $147.94 – $39.99 = $107.95

Part C

5. Answers will vary.

6.

Popular Productions Order Form

Date	Ticket Description	Quantity	Price Each	Price Total
7/17	Adult	4	$20.00	$80.00
	Student	1	15.00	15.00
	Child	3	12.50	37.50
8/3	Senior	10	14.00	140.00
	Adult	2	20.00	40.00

Ticket Total: $312.50
Shipping/Handling: 3.50
Total: $316.00

Unit 4 Review pp. 132–133

Part A

1. 9,753
2. 388
3. 3
4. $150
5. 0
6. 96 or higher
7. 20,046
8. $1.33
9. 24
10. $331.25

11. 13
12. 28 inches
13. 36
14. $17.06
15. 20,244
16. 299
17. 1,277
18. $56.21
19. 40
20. 6

Part B

21. a. 42   56   54   72
    b. 49   63   48   56

22. a. 88   508   330
    b. 48,024   53,792   286,800

23. 1,500   182,000   18,414,000   1,445,760

24. 2,925   85,401   2,762,092   722,680

Part C

25. **53,872 packages**

$1,036 \times 52 = 53,872$

---

26. $11.80

$2 \times $20 = 40

$12 \times $2.35 = 28.20

$40.00 – $28.20 = 11.80

27.

$3 \times 8 = 24$ $4 \times 6 = 24$

28. 425 square feet

$20 \times 20 = 400$

$5 \times 5 = 25$

$400 + 25 = 425$

Working Together

Answers will vary.

Unit 5

When Do I Divide? p. 135

1–2. Answers will vary.

3. divisor 6   quotient 4
4. divisor 7   quotient 8
5. divisor 8   quotient 9
6. dividend 54   divisor 9
7. dividend 28   divisor 4
8. dividend 40   divisor 8

Talk About It

Answers will vary.

Division Strategies pp. 136–137

Part A

1. $24 \div 8 = 3$ $45 \div 9 = 5$ $56 \div 7 = 8$
   $24 \div 3 = 8$ $45 \div 5 = 9$ $56 \div 8 = 7$

2. $48 \div 6 = 8$ $32 \div 4 = 8$ $63 \div 7 = 9$
   $48 \div 8 = 6$ $32 \div 8 = 4$ $63 \div 9 = 7$

Part B

3. $30 \div 5 = 6$
4. $18 \div 3 = 6$

Part C

5. 1   8   1   0
6. 7   0   1   1
7. 1   3   4   1
8. $0 \div 3 = 0$

any number ÷ the same number = 1
$8 \div 1 = 8$

Part D

9. $16 \div 2 = 8$ $56 \div 7 = 8$ $64 \div 8 = 8$
10. $27 \div 3 = 9$ $25 \div 5 = 5$ $48 \div 8 = 6$
11. $0 \div 9 = 0$ $7 \div 7 = 1$ $49 \div 7 = 7$

Division Facts pp. 138–139

Part A

1. a. 3	4	8		b. 2	7	4
2. a. 7	0	8		b. 7	0	1
3. a. 4	2	6		b. 7	7	5
4. a. 3	2	7		b. 2	9	8
5. a. 8	1	3		b. 0	8	1
6. a. 6	5	6		b. 3	0	0
7. a. 8	6	3		b. 1	6	1
8. a. 9	3	9		b. 7	4	4
9. a. 5	9	5		b. 6	6	5
10. a. 7	3	3		b. 4	9	2
11. a. 4	0	9		b. 8	9	2
12. a. 9	5	4		b. 1	10	6
13. a. 4	3	9		b. 7	8	0
14. a. 7	1	9		b. 2	6	5
15. a. 2	2	8		b. 0	5	5

Part B

16. $\frac{20}{4} = 5$

17. $\frac{21}{7} = 3$

18. $\frac{18}{3} = 6$

19. $\frac{54}{6} = 9$

Part C

20. $m = 4$ $t = 7$ $C = 56$
21. $y = 9$ $X = 49$ $r = 8$

Part D

22. $B = 9$ $n = 8$ $W = 7$
23. $A = 8$ $E = 72$ $f = 8$

Dividing by One Digit pp. 140–141

Part A

1. 12   19   121   3,671
2. 131   28   178   1,189

Part B

3. 87   78   65   26   60
4. 1,032   804   1,216   302   491

Part C

5. **22 tables**
   $132 \div 6 = 22$

6. **$640**
   $\$2,560 \div 4 = \$640$

7. **18 teams**
   $162 \div 9 = 18$

8. **61 times as tall**
   $305 \div 5 = 61$

Part D

9. 101   4,907   404   32
10. 107   2,402   3,831   7,015

Remainders pp. 142–143

Part A

1. 8 R1   4 R4   46 R7   94 R1   121 R2
2. 2,663 R1   1,761 R1   426 R4   7,158 R3

Part B

3. **26 bows**
   The answer is 26 R4, which means she can make 26 bows and will have 4 inches of ribbon left over.

4. **Yes**
   $\$100 \div \$6 = 16 \text{ R4}$
   It is enough to buy 16 tickets with $4 left over.

5. **31 jars**
   $245 \div 8 = 30 \text{ R5}$
   The trail mix would fill 30 jars completely, with 5 ounces left over to go in the 31st jar.

Making Connections: Divisibility p. 143

	2	3	4	5	6	8	9	10
73,908	X	X	X		X		X	
1,233		X					X	
2,640	X	X	X	X	X	X		X
1,001								

Zeros in the Answer pp. 144–145

Part A

1. 102   38 R5   1,028 R3   1,112
2. 8,000   201   8,010   900

Part B

3. 5,402   130,207   65,021

4. 5,003   26,668   1,515

Part C

5. 57   62   750   9

6. 400   302   89   190

Making Connections: Estimating Division p. 145

Estimates will vary, but yours should be close to these:

1. 2   100   60

2. 300   60   40

3. **a.** $500   **b.** $400   **c.** $300

Finding an Average pp. 146–147

Part A

1. **7,308 people**

   Step 1: 9,456 + 7,968 + 4,500 = 21,924

   Step 2: 21,924 ÷ 3 = 7,308

2. **20**

   Step 1: 16 + 24 = 40

   Step 2: 40 ÷ 2 = 20

Part B

3. **27°**

   Step 1: 30° + 18° + 22° + 29° + 20° + 36° + 34° = 189°

   Step 2: 189° ÷ 7 = 27°

   On average, snow conditions were ideal that week.

4. **3 inches**

   Step 1: 2 + 4 + 5 + 1 + 3 = 15

   Step 2: 15 ÷ 5 = 3

5. **27°**

   Step 1: 18° + 36° = 54°

   Step 2: 54° ÷ 2 = 27°

6. Friday was not ideal because 36° is 4° over 32°. Saturday was ideal.

Part C

Estimates will vary.

7. **200,000 square feet**

   Step 1: 200 + 125 + 75 + 250 + 100 + 175 + 375 + 300 = 1,600 thousand square feet

   Step 2: 1,600 thousand ÷ 8 = 200 thousand square feet = 200,000 square feet

8. **225,000 square feet; five stores**

   Step 1: 75 thousand + 375 thousand = 450 thousand

   Step 2: 450 thousand ÷ 2 = 225 thousand = 225,000

9. **50,000 square feet**

   1,000,000 − 200,000 = 800,000

   800,000 ÷ 16 = 50,000

Part D

Milwaukee Brewers

Player	At Bats	Hits	1,000 at Bats	
Vaughn	70	20	20 × 1,000 = 20,000	20,000 ÷ 70 = 285
Spiers	260	90	90 × 1,000 = 90,000	90,000 ÷ 260 = 346
Diaz	410	90	90 × 1,000 = 90,000	90,000 ÷ 410 = 219
Jaha	320	80	80 × 1,000 = 80,000	80,000 ÷ 320 = 250

10. **a.** Spiers

    b. Diaz

11. Answers will vary.

12. Answers will vary. Players need to get more hits in fewer at bats.

Mixed Review pp. 148–149

Part A

1. 1   1   6   0   9

2. 3   8   4   4   3

3. 3   6   9   4   9

4. 7   6   3   5   6

5. 6   5   5   2   9

6. 9   0   11   5   7

Part B

7. 11   12   129   1,908   73

8. 805   306   2,169 R2   4,902   1,006

9. 4 R5   187 R2   7,050 R4   1,268 R2   1,258 R2

10. 24   78   96

11. 42   90   567   201

12. 46 R70   44 R300   81

Part C

13. **$126**

    $756 ÷ 6 = $126

14. $N = 4$ $t = 10$ $z = 49$ b = any number

15. 77 R2   OK   806 R3

16. a. **30 miles per gallon**

   $300 \div 10 = 30$

   b. **4 trips to and from work**

   $300 \div 70 = 4\ R20$

17. a. **40 posts**

   $200 \div 5 = 40$

   b. **4 bundles**

   $3 \times 12 = 36$ posts, too few; $4 \times 12 = 48$ posts

18. **$60,000**

   Step 1: $12,000,000 \div 20 = \$600,000$ each year

   Step 2: $\$600,000 \div 10 = \$60,000$ per person

Part D

19. 3   4 *or* 8   10

20. 9   5   6

21. 2   7   8 *or* 4

22. Answers will vary.

Dividing by Two or More Digits pp. 150–151

Part A

1. 2   15   12   20   12 R40

2. 21   19   4 R12   10 R20   85 R82

3. 216 R50   643   2,090 R32   1,171

Part B

4. 48   22   52   405

5. 403 R2   906   108   302

Part C

Item Sets pp. 152–153

Part A

1. Information needed: all wages, 5 people

   Calculation: $320 + $480 + $900 + $350 + $150 = $2,200

   $2,200 \div 5 = $**440**

2. Information needed: $900, 60 hours

   Calculation: $900 \div 60 =$ **$15 per hour**

3. Answers will vary.

Part B

4.

	Mileage Log			
Day	Beginning Odometer	Ending Odometer	Miles Driven	Gallons of Gas
Monday	21,345	21,555	210	7
Tuesday	21,555	21,735	180	6
Wednesday	21,735	21,855	120	4
Thursday	21,855	22,095	240	8
Friday	22,095	22,185	90	3
		Total:	840	

5. **840 miles**

   $210 + 180 + 120 + 240 + 90 = 840$

6. **30 miles per gallon; answers will vary.**

   $840 \div 28 = 30$

7. **168 miles**

   $840 \div 5 = 168$

Part C

8. **90 percent. They have a high percentage of calories from fat.**

   $3 \times 900 = 2,700$

   $2,700 \div 30 = 90$

   $90\% > 30\%$

9. **23 percent. They have a low percentage of calories from fat.**

   $6 \times 900 = 5,400$

   $5,400 \div 230 = 23\ R110$

   $23\% < 30\%$

10. **42 percent. They have a high percentage of calories from fat.**

   $10 \times 900 = 9,000$

   $9,000 \div 210 = 42\ R180$

   $42\% > 30\%$

Dividing Dollars and Cents pp. 154–155

Part A

1. $1.15   2. $37.06

   $5.83 $217

   $7.21 $.50

3. $.04

$.06

$.18

4. $.08

$.46

$4.80

Part B

5. $.16

$3.45

$834.12

7. $.07

$1.37

$.05

6. $48.50

$61.24

$50

Part C

8. $34.50

$552 ÷ 16 = $34.50

9. $36.54

Step 1: $48.72 × 3 = $146.16

Step 2: $146.16 ÷ 4 = $36.54

10. $123.06

Step 1: $48.75 + $24.36 + $36.48 + $19.69 + $184.39 + $42.53 + $12.98 = $369.18

Step 2: $369.18 ÷ 3 = $123.06

Part D

11. 386 miles

Step 1: 250 + 175 + 84 + 320 + 1,026 + 290 + 95 + 848 = 3,088

Step 2: 3,088 ÷ 8 = 386

12. Family G will spend $15 per person, and family H will spend $25, a difference of $10 per person.

13. $37 per person

$48 for hotel room

$25 per person for food

$25 × 4 = $100 per day for food

$100 + $48 = $148 per day for food and lodging

$148 ÷ 4 = $37 per person

14. Answers will vary.

Dividing on a Calculator pp. 156–157

Part A

1. Divisor : 92   Answer: 2

Divisor: 16   Answer: 105

Divisor: 25   Answer: 123

2. Divisor: 48   Answer: 3,215

3. Divisor: 395   Answer: 57

4. Divisor: 9   Answer: 2,856

Part B

Estimates will vary.

5. Estimate: 400   Estimate: 300   Estimate: 300

Answer: 485   Answer: 340   Answer: 259

6. Estimate: $25   Estimate: $50   Estimate: $200

Answer: $25.08   Answer: $54.50   Answer: $246.03

Part C

7. 52 R8

674 R7

337 R51

Making Connections: Cutting the Cake p. 157

Answers will vary.

Choosing the Right Operation pp. 158–159

Part A

Answers will vary.

Part B

1. addition

2. division

3. subtraction and addition

Part C

4. 4 gallons

Step 1: 672 × 2 = 1,344

Step 2: 1,344 ÷ 400 = 3 R144

5. $15.85 per square yard

Step 1: $12.95 + $18.95 + $16.50 + $15 = $63.40

Step 2: $63.40 ÷ 4 = $15.85

6. 336 square feet

Step 1: 6 × 8 = 48

Step 2: 12 × 24 = 288

Step 3: 48 + 288 = 336

7. 38 square yards

336 ÷ 9 = 37 R3

8. $666.96

Step 1: 4 × $9.99 = $39.96

Step 2: 38 × $16.50 = $627

Step 3: $39.96 + $627 = $666.96

Part D

9. **800 calories**

    $120 \div 15 = 8$

    $8 \times 100 = 800$

10. **120 calories**

    $110 + 408 + 320 = 838$

    $838 \div 7 = 119 \text{ R5} \approx 120$

11. Answers will vary.

    Running and cross-country skiing would burn the most calories.

Finding Unit Price pp. 160–161

Part A

1. K-Pops: $.15

2. Oat-Os: $.16

3. Wheat Treats: $.13

    Best buy: Wheat Treats

Part B

4. **$6.72 per cassette**

    Step 1: $8.96 \times 3 = $26.88

    Step 2: $26.88 \div 4$ cassettes = $6.72

5. **$.15 per pound**

    $1.50 \div 10$ pounds = $.15

6. **$.41 per bottle**

    $3.28 \div 8$ bottles = $.41 per bottle

7. **$17.24 per pair of shoes**

    Step 1: $22.98 \div 2 = $11.49

    Step 2: $22.98 + $11.49 = $34.47

    Step 3: $34.47 \div 2$ pairs of shoes = $17.24

Part C

8.

Inventory

Item	Cost	Unit Price	Selling Price
Quality T-shirt	$49.92/doz.	$4.16	$8.32
Long sleeve T-shirt	$93.60/doz.	7.80	15.60
Baseball caps	$156/case	6.50	13.00
Golf shirts/pocket	$57.90/6-pack	9.65	19.30
Heavyweight sweatshirt	$237.80/doz.	19.82	39.64
Jersey shorts	$49.80/6-pack	8.30	16.60

9. $18.75

10. Answers will vary.

Part D

11. $17.99 \div 40 = 0.44 R.39 \approx **$0.45 per pound** (best deal)

    $5.99 \div 8 = 0.74 R.07 \approx **$0.75 per pound**

12. $2.89 \div 4 = 0.72 R.01 \approx **$0.73 per quart** (best deal)

    $89\cent = **$.89 per quart**$

    $1.64 \div 2 = **$.82 per quart**

13. $5.98 \div 12 = 0.49 R.10 \approx **$.50 per pint** (best deal)

    $79\cent = **$.79 per pint**$

14. Answers will vary.

Putting It All Together pp. 162–163

Part A

Estimates will vary.

Estimate: $100

Exact: $108.69

Step 1: $90.81 + $110.12 + $128.60 + $105.23 = $434.76

Step 2: $434.76 \div 4 = $108.69

Part B

Estimates will vary.

Estimate: $460

Exact: $505.76

Step 1: $40 \times $8.72 = $348.80

$12 \times $13.08 = $156.96

Step 2: $348.80 + $156.96 = $505.76

Part C

144 tiles

9 ft. $\times$ 16 ft. = 144 sq. ft.

47 feet of baseboard

Step 1: $9 + 9 + 16 + 16 = 50$

Step 2: $50 - 3 = 47$

Part D

Answers will vary.

Part E

1. **$375**

    $750 - $75 = $675

    $675 - $300 = $375

2. **$494**

    $375 - $30 = $345

    $345 \div 3 = $115

    $345 - $115 = $230

    $724 - $230 = $494

3. **$869**

    $375 + $494 = $869

4. Answers will vary.

5. Answers will vary.

Part F

$n = 84$

- $n < 100$: The number is between 0 and 99
- n is even: The number must end in 0, 2, 4, 6, or 8.
- $n > 56$: n is between 58 and 98.
- The sum of the digits is 12: either 66 or 84.
- n is evenly divisible by 7: $84 \div 7 = 12$

Unit 5 Review pp. 164–165

Part A

1. 6,000

2. any number less than 312

3. 10:45

4. $T = 8$

5. 317

6. 32 inches

7. 9

    The pattern is $- 3; + 5$

8. $1.20

9. $27.65

10. $m = 48$

11. 2,144

12. $R =$ any number

13. $F = 74$

14. 72 square feet

15. 3,400

16. $n = 9$

17. 2,880

Part B

18.	7	5	3	6	6
19.	9	8	7	4	9
20.	25	240	8	10	740
21.	1,049	42 R2	23	69 R33	40
22.	117 R70	2,001	21	5 R5	

Part C

23. **60 bricks**

    Step 1: $12 + 8 + 12 + 8 = 40$ feet

    Step 2: 40 feet $= 480$ inches

    $480 \div 8 = 60$

24. **$506.85**

    Step 1: $1,308 \div 12 = 109$ dozen

    Step 2: $109 \times \$4.65 = \$506.85$

25. Divisible by 1, 2, 3, 4, 5, 6, 8, 9, 10

26. $H = 420$

    $W =$ any number

    $x = 64$

Part D

27. **21 pages per day**

    $294 \div 14 = 21$

28. **$24**

    Step 1: $64 + 48 + 28 + 30 = 170$

    Step 2: $\$4,080 \div 170 = \$24$

29. **$37.50**

    Step 1: $\$27.75 \div 37 = \$.75$

    Step 2: $\$.75 \times 50 = \$37.50$

30. Answers will vary.

Working Together

Answers will vary.

Glossary

addition the operation that combines amounts (p. 42)

$$5 + 2 = 7$$

XXXXX + XX = XXXXXXX

area surface measurement in square units. The area of a rectangle is length times width (p. 122).

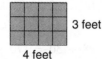

3 feet

4 feet

Area = 3 feet × 4 feet = 12 square feet

average the number that best represents a group of numbers. To find the average, add a group of numbers, then divide by the number of values in the group (p. 146).

The ages of the parents of the preschool class are 27 years, 32 years, 40 years, 35 years, 34 years, and 30 years.

$$27 + 32 + 40 + 35 + 34 + 30 = 198$$
$$198 \div 6 = 33$$

The average age is 33 years old.

bar graph a picture that uses bars to compare values (p. 119)

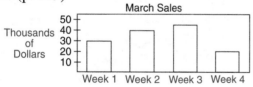

borrow to regroup from a larger place value in subtraction (p. 80)

$$\begin{array}{r} {}^{4}\cancel{5}{}^{1}7 \\ -\ 28 \\ \hline 29 \end{array}$$

calendar a chart that shows the months and days of a year (p. 28)

FEBRUARY						
S	M	T	W	T	F	S
	1	2	3	4	5	6	7
8	9	10	11	12	13	14
15	16	17	18	19	20	21
22	23	24	25	26	27	28

carry to regroup to a larger place value in addition and multiplication (p. 52)

$$\begin{array}{r} {}^{1}\ \\ 35 \\ +\ 26 \\ \hline 61 \end{array}$$

change the difference between what you pay and what you owe (p. 32)

You use a $20 bill to pay for a purchase of $18.75.

$$\begin{array}{r} \$20.00 \\ -\ 18.75 \\ \hline \$1.25 \end{array}$$ is your change.

circle graph a picture that uses a circle to show how a whole amount is divided into parts; also known as a pie chart (p. 115)

cube a box whose sides have equal length, width, and height. Each side is a square (p. 123).

cubic unit a unit used to measure volume, such as cubic foot or cubic centimeter (p. 123)

decimal point a point that separates whole numbers from decimals and dollars from cents. Numbers written after the decimal point indicate cents (p. 31).

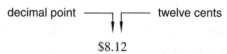

decimal point ⎯⎯⎯ twelve cents

$8.12

deductions amount subtracted from earnings. Deductions include state and federal taxes and Social Security taxes (p. 92).

You earn $540. Federal tax is $97, state tax is $11, and Social Security is $38.

$$\$97 + \$11 + \$38 = \$146 \text{ total deductions}$$
$$\$540 - \$146 = \$394 \text{ take-home pay}$$

difference the answer to a subtraction problem (p. 70)

$$22 - 10 = \underset{\uparrow}{12} \quad \text{difference}$$

digit a symbol for a number. The digits are 0, 1, 2, 3, 4, 5, 6, 7, 8, and 9 (p. 16).

dividend in a division problem, the number being divided (p. 135)

$$\underset{\uparrow}{39} \div 13 = 3 \quad \text{dividend}$$

divisible the description of a number that can be evenly divided with no remainder (p. 143)

18 is evenly divisible by 6 because
$18 \div 6 = 3$ with no remainder

division the operation that separates an amount into equal parts *or* finds how many times one amount fits into another (p. 134)

Example 1: Divide a 24-inch board into 3 equal pieces.

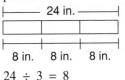

$24 \div 3 = 8$
Each part is 8 inches.

Example 2: How many 3-inch pieces can be cut from a 24-inch board?

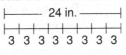

$24 \div 3 = 8$
There are 8 pieces.

division symbols symbols used to show division. For example, different symbols can be used to show 15 divided by 3 (p. 135):

$$15 \div 3 \ or \ 3\overline{)15} \ or \ \tfrac{15}{3}$$

divisor in a division problem, the number that an amount is being divided by (p. 135)

$$28 \div \underset{\uparrow}{4} = 7 \quad \text{divisor}$$

equal sign the symbol (=) that shows that two amounts have the same value (p. 18)

$$9 + 5 \underset{\uparrow}{=} 2 \times 7 \quad \text{means "is equal to"}$$

equation a number sentence that contains an equal sign (p. 49)

$$8 \div 2 = 4$$

This equation is read as "Eight divided by two equals four."

estimate an answer that is reasonably close to the exact answer (p. 23)

exact	estimate
26	26 ≈ 30
+ 42	+ 42 ≈ + 40
68	70

even number any whole number whose last digit is 0, 2, 4, 6, or 8 (p. 17)

factors numbers that are multiplied (p. 104)

$$7 \times 8 = 56 \qquad 7 \text{ and } 8 \text{ are the factors of } 56.$$

geometry the study of shapes (p. 18)

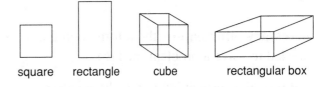

square   rectangle   cube   rectangular box

greater than the symbol (>) that shows one amount is larger than another (p. 18)

$$5 \underset{\uparrow}{>} 3 \quad \text{means "is greater than"}$$

gross pay total earnings before deductions (p. 92)

less than the symbol (<) that shows one amount is smaller than another (p. 18)

$$3 \underset{\uparrow}{<} 5 \quad \text{means "is less than"}$$

line a straight geometric figure that has unending length (p. 18)

The arrows show that the line is unending.

measurement a unit that shows length, capacity, time, money, area, or volume (p. 19). Refer to the table of measures on page 206.

minus sign the symbol (−) that shows subtraction (p. 71)

means "minus"

$$12 - 9 = 3$$

multiplication the operation that combines like amounts (p. 100)

$$5 \times 2 = 10$$

multiplier a number that multiplies another (p. 104)

multiplier

$$8 \times 5 = 40$$

multistep problem a math problem that requires more than one operation to solve (p. 128)

necessary information the information that is needed to solve a problem (p. 116)

net pay the amount left to take home after deductions are subtracted from gross pay (p. 92)

You earn $400 and your deductions are $100. Your net pay is $400 − $100, or $300.

number line a line that shows the order of numbers (p. 18)

0   10   20   30   40   50

number sense knowing what a number means and how to use it (p. 12)

odd number any whole number whose last digit is 1, 3, 5, 7, or 9 (p. 17)

partial product the result from each digit when you multiply by a number with two or more digits (p. 114)

$$\begin{array}{r} 52 \\ \times\ 24 \\ \hline 208 \\ 1\ 04 \\ \hline 1{,}248 \end{array}$$

partial products

pattern a sequence of numbers or items that continue according to a rule (p. 16)

2, 5, 8, 11, 14, . . .
This pattern follows the rule of adding 3 to get the next value.

perimeter the distance around a geometric shape. To find the perimeter, add the lengths of all sides of the figure (p. 55).

3 in.   5 in.
2 in.   4 in.
8 in.
$$P = 2 + 3 + 5 + 4 + 8 = 22 \text{ inches}$$

place value the value of a digit that depends on its position in the number (p. 20)

83   The 8 has a value of 80 because it is in the tens place.

803   The 8 has a value of 800 because it is in the hundreds place.

plus sign the symbol (+) that shows addition (p. 42)

means "added to"

$$5 + 3 = 8$$

polygon a flat, closed geometric figure with straight sides (p. 54). Some examples are:

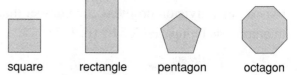

square   rectangle   pentagon   octagon

product the answer to a multiplication problem (p. 104)

product

$$4 \times 5 = 20$$

quotient the answer to a division problem (p. 135)

$$\begin{array}{r} 4 \\ 5\overline{)20} \end{array}$$ 4 is the quotient.

rectangle a four-sided polygon with opposite sides equal and four right angles (p. 54)

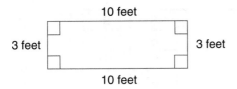

regroup to borrow an amount from a larger place value in subtraction *or* to carry an amount to a larger place value in addition and multiplication (p. 52)

$$\begin{array}{r} \overset{4}{\cancel{5}}\overset{1}{3} \\ -\ 2\,5 \\ \hline 2\,8 \end{array} \qquad \begin{array}{r} \overset{1}{38} \\ +\ 2\,7 \\ \hline 6\,5 \end{array} \qquad \begin{array}{r} \overset{2}{15} \\ \times\ 5 \\ \hline 7\,5 \end{array}$$

remainder the amount left over when a number doesn't divide evenly (p. 142)

$$\begin{array}{r} 7\ R3 \\ 5\overline{)38} \\ -\ 35 \\ \hline 3 \end{array}$$ 7 remainder 3

rounding approximating the value of a number to make it easier to work with (p. 22)

Round 285 to the nearest 100.
$285 \approx 300$

└── means "is approximately equal to"

square a four-sided polygon with equal sides and four right angles (p. 54)

square unit a unit used to measure area, such as a square inch or a square centimeter (p. 122)

subtraction the operation used to take away an amount to get a smaller amount *or* to find the difference between two numbers to make a comparison (p. 70)

Example 1   Example 2

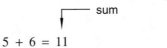

$$\begin{array}{r} \$40 \\ -\ 15 \\ \hline \$25 \end{array} \qquad \begin{array}{r} 8\ \text{feet} \\ -\ 3\ \text{feet} \\ \hline 5\ \text{feet longer} \end{array}$$

sum the answer to an addition problem (p. 43)

$$5 + 6 = 11$$ sum

tally system a method of counting and marking numbers in groups of five (p. 16)

times sign the symbol (×) that shows multiplication (p. 100)

unit price the price of an item divided by the number of units (p. 160)

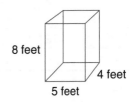

$$\begin{array}{r} \$\ .09 \\ 50\overline{)\$4.50} \end{array}$$ Each tablet costs 9 cents.

variable a symbol used to represent an unknown value. Letters are often used and have different values in different problems (p. 49).

$n + 5 = 8$ *n* is the variable.

volume measurement in cubic units. The volume of a rectangular box is length times width times height (p. 123).

$V = 4 \times 5 \times 8 = 160$ cubic feet

Tool Kit

The Addition Table

sum

The place where a row and a column meet is the **sum.** Example: $3 + 5 = 8$

+	0	1	2	3	4	5	6	7	8	9
0	0	1	2	3	4	5	6	7	8	9
1	1	2	3	4	5	6	7	8	9	10
2	2	3	4	5	6	7	8	9	10	11
3	3	4	5	6	7	8	9	10	11	12
4	4	5	6	7	8	9	10	11	12	13
5	5	6	7	8	9	10	11	12	13	14
6	6	7	8	9	10	11	12	13	14	15
7	7	8	9	10	11	12	13	14	15	16
8	8	9	10	11	12	13	14	15	16	17
9	9	10	11	12	13	14	15	16	17	18

You can also use the addition table to find subtraction facts. Given the fact that $3 + 5 = 8$, the related subtraction facts are $8 - 5 = 3$ and $8 - 3 = 5$.

The Multiplication Table

product

The place where a row and a column meet is the **product.** Example: $3 \times 5 = 15$

×	0	1	2	3	4	5	6	7	8	9	10
0	0	0	0	0	0	0	0	0	0	0	0
1	0	1	2	3	4	5	6	7	8	9	10
2	0	2	4	6	8	10	12	14	16	18	20
3	0	3	6	9	12	15	18	21	24	27	30
4	0	4	8	12	16	20	24	28	32	36	40
5	0	5	10	15	20	25	30	35	40	45	50
6	0	6	12	18	24	30	36	42	48	54	60
7	0	7	14	21	28	35	42	49	56	63	70
8	0	8	16	24	32	40	48	56	64	72	80
9	0	9	18	27	36	45	54	63	72	81	90
10	0	10	20	30	40	50	60	70	80	90	100

You can also use the multiplication table to find division facts. Given the fact that $3 \times 5 = 15$, the related division facts are $15 \div 5 = 3$ and $15 \div 3 = 5$.

Tool Kit

Common Symbols and Formulas

You can use the number line to help you picture the examples.

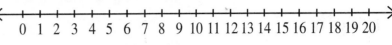

Symbol	Meaning	Examples	
+	plus; added to	$4 + 5 = 9$	
−	minus	$10 - 6 = 4$	
×	times; multiplied by	$3 \times 4 = 12$	
÷	divided by	$10 \div 2 = 5$	
<	is less than	$4 < 7$	$3 + 2 < 7 + 1$
>	is greater than	$6 > 5$	$4 + 4 > 5 + 1$
=	is equal to	$3 = 3$	$2 + 4 = 1 + 5$
≈	is approximately equal to	$12 \approx 10$	$17 \approx 20$

Formulas for Rectangles and Squares	Meaning	Examples
$P = l + w + l + w$ $\quad = 2l + 2w$	Perimeter = length + width + length + width	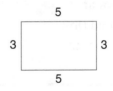 $P = 5 + 3 + 5 + 3 = 16$
	or $P = 2 \times$ length $+ 2 \times$ width	*or* $P = 2 \times 5 + 2 \times 3$ $\quad = 10 + 6$ $\quad = 16$
$A = l \times w$	Area = length $\times$ width	$A = 5 \times 3 = 15$
$V = l \times w \times h$	Volume = length $\times$ width $\times$ height	$V = 4 \times 2 \times 3$ $\quad = 8 \times 3$ $\quad = 24$

Tool Kit

Common Measurements

Length

English Measurement

12 inches = 1 foot (ft.)
3 feet = 1 yard (yd.)
1,760 yards = 1 mile (mi.)
5,280 feet = 1 mile

Metric Measurement

10 millimeters (mm) = 1 centimeter (cm)
100 centimeters = 1 meter (m)
1,000 meters = 1 kilometer (km)

Capacity

English Measurement

8 fluid ounces (fl. oz.) = 1 cup (c.)
2 cups = 1 pint (pt.)
2 pints = 1 quart (qt.)
4 quarts = 1 gallon (gal.)

Metric Measurement

1,000 milliliters (ml) = 1 liter
1,000 liters = 1 kiloliter (kl)

Weight

English Measurement

16 ounces = 1 pound (lb.)
2,000 pounds = 1 ton (tn.)

Metric Measurement

1,000 milligrams (mg) = 1 gram (g)
1,000 grams = 1 kilogram (kg)
1,000 kilograms = 1 metric ton (t)

Calendar Time

January (Jan.) 31 days
February (Feb.) 28 days (29 in leap year)
March (Mar.) 31 days
April (Apr.) 30 days

May 31 days
June 30 days
July 31 days
August (Aug.) 31 days

September (Sept.) 30 days
October (Oct.) 31 days
November (Nov.) 30 days
December (Dec.) 31 days

U.S. Money

1 penny ($.01)

1 quarter ($.25) = 5 nickels

1 nickel ($.05) = 5 pennies

1 dollar ($1.00) = 4 quarters

1 dime ($.10) = 2 nickels

Tool Kit

Measurement Tools

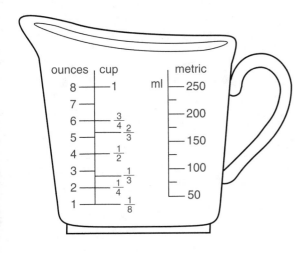

A **measuring cup** is used to measure liquid and granular amounts. Measuring cups usually have either English measurements (cups and ounces) or metric measurements (milliliters). The measuring cup on the left shows both English and metric measurements.

Scales are used to measure weight. Scales usually use either English measurements (ounces or pounds) or metric measurements (grams or kilograms). The scale on the right measures pounds.

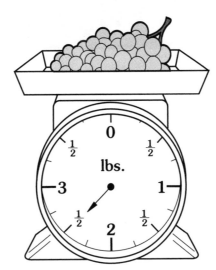

A **ruler** is used to measure length. An English ruler measures in inches. A metric ruler measures in centimeters (cm) and millimeters (mm). The ruler on the right shows both English and metric measurements.

DATE DUE

Demco, Inc. 38-293